W0018135

Chemical-Mechanical Polishing 2000— Fundamentals and Materials Issues

MATERIALS RESEARCH SOCIETY
SYMPOSIUM PROCEEDINGS VOLUME 613

Chemical-Mechanical Polishing 2000— Fundamentals and Materials Issues

Symposium held April 26–27, 2000, San Francisco, California, U.S.A.

EDITORS:

Rajiv K. Singh
University of Florida
Gainesville, Florida, U.S.A.

Rajeev Bajaj
Applied Materials
Santa Clara, California, U.S.A.

Mansour Moinpour
Intel Corporation
Santa Clara, California, U.S.A.

Marc Meuris
IMEC VZW
Leuven, Belgium

Materials Research Society
Warrendale, Pennsylvania

Single article reprints from this publication are available through
University Microfilms Inc., 300 North Zeeb Road, Ann Arbor, Michigan 48106

CODEN: MRSPDH

Published by:

Materials Research Society
506 Keystone Drive
Warrendale, PA   15086
Telephone (724) 779-3003
Fax (724) 779-8313
Web site: http://www.mrs.org/

Library of Congress Cataloging-in-Publication Data

Manufactured in the United States of America

CONTENTS

CMP CONSUMABLES

PROCESS INTEGRATION AND
MANUFACTURABILITY

*Invited Paper

PREFACE

This proceedings volume contains papers from Symposium E, "Fundamentals and Materials Issues in Chemical-Mechanical Polishing of Materials," held April 26–27 at the 2000 MRS Spring Meeting in San Francisco, California. Significant interest and advancements in the field were represented in more than 55 oral and poster presentations. Papers both in the application and fundamentals of CMP were presented at this conference. With the rapid emergence of copper and low dielectric constant materials, the interest in CMP is expected to further accelerate in future years.

We would like to thank all the authors, particularly the invited speakers, the session chairs and referees, all who helped to make this symposium a success. We would like to especially thank Ara Philipossian (University of Arizona), Ajoy Zutshi (Applied Materials), R. Tiwari (Texas Instruments), and S. Das (IBM) for helping us review the manuscripts. It is also our pleasure to acknowledge the support of the Department of Materials Science and Engineering and the Engineering Research Center for Particle Science and Technology at the University of Florida for financial support to this program.

Rajiv K. Singh
Rajeev Bajaj
Mansour Moinpour
Marc Meuris

February 2001

MATERIALS RESEARCH SOCIETY SYMPOSIUM PROCEEDINGS

Volume 578— Multiscale Phenomena in Materials—Experiments and Modeling, I.M. Robertson, D.H. Lassila, R. Phillips, B. Devincre, 2000, ISBN: 1-55899-486-6

Volume 579— The Optical Properties of Materials, J.R. Chelikowsky, S.G. Louie, G. Martinez, E.L. Shirley, 2000, ISBN: 1-55899-487-4

Volume 580— Nucleation and Growth Processes in Materials, A. Gonis, P.E.A. Turchi, A.J. Ardell, 2000, ISBN: 1-55899-488-2

Volume 581— Nanophase and Nanocomposite Materials III, S. Komarneni, J.C. Parker, H. Hahn, 2000, ISBN: 1-55899-489-0

Volume 582— Molecular Electronics, S.T. Pantelides, M.A. Reed, J. Murday, A. Aviram, 2000, ISBN: 1-55899-490-4

Volume 583— Self-Organized Processes in Semiconductor Alloys, A. Mascarenhas, D. Follstaedt, T. Suzuki, B. Joyce, 2000, ISBN: 1-55899-491-2

Volume 584— Materials Issues and Modeling for Device Nanofabrication, L. Merhari, L.T. Wille, K.E. Gonsalves, M.F. Gyure, S. Matsui, L.J. Whitman, 2000, ISBN: 1-55899-492-0

Volume 585— Fundamental Mechanisms of Low-Energy-Beam-Modified Surface Growth and Processing, S. Moss, E.H. Chason, B.H. Cooper, T. Diaz de la Rubia, J.M.E. Harper, R. Murti, 2000, ISBN: 1-55899-493-9

Volume 586— Interfacial Engineering for Optimized Properties II, C.B. Carter, E.L. Hall, S.R. Nutt, C.L. Briant, 2000, ISBN: 1-55899-494-7

Volume 587— Substrate Engineering—Paving the Way to Epitaxy, D. Norton, D. Schlom, N. Newman, D. Matthiesen, 2000, ISBN: 1-55899-495-5

Volume 588— Optical Microstructural Characterization of Semiconductors, M.S. Unlu, J. Piqueras, N.M. Kalkhoran, T. Sekiguchi, 2000, ISBN: 1-55899-496-3

Volume 589— Advances in Materials Problem Solving with the Electron Microscope, J. Bentley, U. Dahmen, C. Allen, I. Petrov, 2000, ISBN: 1-55899-497-1

Volume 590— Applications of Synchrotron Radiation Techniques to Materials Science V, S.R. Stock, S.M. Mini, D.L. Perry, 2000, ISBN: 1-55899-498-X

Volume 591— Nondestructive Methods for Materials Characterization, G.Y. Baaklini, N. Meyendorf, T.E. Matikas, R.S. Gilmore, 2000, ISBN: 1-55899-499-8

Volume 592— Structure and Electronic Properties of Ultrathin Dielectric Films on Silicon and Related Structures, D.A. Buchanan, A.H. Edwards, H.J. von Bardeleben, T. Hattori, 2000, ISBN: 1-55899-500-5

Volume 593— Amorphous and Nanostructured Carbon, J.P. Sullivan, J. Robertson, O. Zhou, T.B. Allen, B.F. Coll, 2000, ISBN: 1-55899-501-3

Volume 594— Thin Films—Stresses and Mechanical Properties VIII, R. Vinci, O. Kraft, N. Moody, P. Besser, E. Shaffer II, 2000, ISBN: 1-55899-502-1

Volume 595— GaN and Related Alloys—1999, T.H. Myers, R.M. Feenstra, M.S. Shur, H. Amano, 2000, ISBN: 1-55899-503-X

Volume 596— Ferroelectric Thin Films VIII, R.W. Schwartz, P.C. McIntyre, Y. Miyasaka, S.R. Summerfelt, D. Wouters, 2000, ISBN: 1-55899-504-8

Volume 597— Thin Films for Optical Waveguide Devices and Materials for Optical Limiting, K. Nashimoto, R. Pachter, B.W. Wessels, J. Shmulovich, A.K-Y. Jen, K. Lewis, R. Sutherland, J.W. Perry, 2000, ISBN: 1-55899-505-6

Volume 598— Electrical, Optical, and Magnetic Properties of Organic Solid-State Materials V, S. Ermer, J.R. Reynolds, J.W. Perry, A.K-Y. Jen, Z. Bao, 2000, ISBN: 1-55899-506-4

Volume 599— Mineralization in Natural and Synthetic Biomaterials, P. Li, P. Calvert, T. Kokubo, R.J. Levy, C. Scheid, 2000, ISBN: 1-55899-507-2

Volume 600— Electroactive Polymers (EAP), Q.M. Zhang, T. Furukawa, Y. Bar-Cohen, J. Scheinbeim, 2000, ISBN: 1-55899-508-0

Volume 601— Superplasticity—Current Status and Future Potential, P.B. Berbon, M.Z. Berbon, T. Sakuma, T.G. Langdon, 2000, ISBN: 1-55899-509-9

Volume 602— Magnetoresistive Oxides and Related Materials, M. Rzchowski, M. Kawasaki, A.J. Millis, M. Rajeswari, S. von Molnár, 2000, ISBN: 1-55899-510-2

Volume 603— Materials Issues for Tunable RF and Microwave Devices, Q. Jia, F.A. Miranda, D.E. Oates, X. Xi, 2000, ISBN: 1-55899-511-0

Volume 604— Materials for Smart Systems III, M. Wun-Fogle, K. Uchino, Y. Ito, R. Gotthardt, 2000, ISBN: 1-55899-512-9

MATERIALS RESEARCH SOCIETY SYMPOSIUM PROCEEDINGS

CMP Mechanisms

Mat. Res. Soc. Symp. Vol. 613 © 2000 Materials Research Society

The Effect of Wafer Shape on Slurry Film Thickness and Friction Coefficients in Chemical Mechanical Planarization

Joseph Lu[a], Jonathan Coppeta[a], Chris Rogers[a], Vincent P. Manno[a], Livia Racz[a], Ara Philipossian[b], Mansour Moinpour[b], Frank Kaufman[c]
[a]Tufts University, Dept. Mechanical Engineering, Medford, MA 02155 USA
[b]Intel Corporation, Santa Clara, CA 95052 USA
[c]Cabot Corporation, Aurora, IL 60504 USA

ABSTRACT

The fluid film thickness and drag during chemical-mechanical polishing are largely dependent on the shape of the wafer polished. In this study we use dual emission laser induced fluorescence to measure the film thickness and a strain gage, mounted on the polishing table, to measure the friction force between the wafer and the pad. All measurements are taken during real polishing processes. The trends indicate that with a convex wafer in contact with the polishing pad, the slurry layer increases with increasing platen speed and decreases with increasing downforce. The drag force decreases with increasing platen speed and increases with increasing downforce. These similarities are observed for both in-situ and ex-situ conditioning. However, these trends are significantly different for the case of a concave wafer in contact with the polishing pad. During ex-situ conditioning the trends are similar as with a convex wafer. However, in-situ conditioning decreases the slurry film layer with increasing platen speed, and increases it with increasing downforce in the case of the concave wafer. The drag force increases with increasing platen speed as well as increasing downforce. Since we are continually polishing, the wafer shape does change over the course of each experiment causing a larger error in repeatability than the measurement error itself. Different wafers are used throughout the experiment and the results are consistent with the variance of the wafer shape. Local pressure measurements on the rotating wafer help explain the variances in fluid film thickness and friction during polishing.

INTRODUCTION

Chemical mechanical planarization (CMP) is widely used in the manufacturing process of very large scale integrated (VLSI) circuits and ultra large scale integrated (ULSI) circuits. Some applications of these circuits include processor chips, RAM chips, and hard drives. The advantage of the CMP process is that both local and global planarity can be achieved. Planarity on the die level, and wafer level is even more desirable as production moves to 200 mm and 300 mm wafers. Both local and global planarity are extremely important in multilevel circuits with feature sizes smaller than 500 nm. The fine depth of focus requirements in current and future optical lithography techniques put extreme demands on the processes involved. The increased control of the CMP process allows for more efficient use of resources, and will aid in the development of future nanoscale circuit topographies.

The CMP process is widely used, yet there is only a limited understanding of the fundamental mechanisms involved. Fundamental research has been done to both experimentally understand polishing characteristics and analytically model the process involved [1,2,3]. Slurry film thickness during polishing is largely determined by how the hydrodynamic pressure and pad aesperity contact pressure equilibrate with the downforce applied. Some researchers have found that during polishing there is a vacuum created underneath the wafer and a negative vertical pad displacement [4,5,6]. Others have found positive fluid pressure developing in the gap between the wafer and polishing pad [7]. In our experiments we have witnessed both cases. A concave wafer displays the vacuum effect, and a convex wafer develops positive fluid pressure in the gap. Frictional characteristics during polishing will depend on whether there is positive or negative (suction) pressure underneath the wafer. The frictional information can be used to determine what lubrication regime is involved [8]. Cook and Su have established that removal rate is largely dependent on the lubrication regime [9, 10]. Fluid thickness and polishing performance also vary in different lubrication regimes [11]. Sundarajann, Cook and Su have all cited that wafer curvature might be a factor in determining the lubrication regime involved, but little work has been done to study this effect partly due to the fact the wafer shape is difficult to control. Some numerical models analytically solve for the fluid depth between the wafer and pad as well as the wafer angle of attack by assuming a natural bow to the wafer [12].

In this paper, we will investigate the specific effect wafer curvature has on slurry film thickness and friction during CMP. Experiments have been conducted to examine how the slurry fluid thickness and friction changes with wafer shape and polishing parameters, such as platen speed and wafer downforce.

EXPERIMENT

Figure 1 shows the modified rotary polisher used to study slurry flow beneath the wafer. We use a Struers RotoPol-31 tabletop polisher to rotate a 300 mm (12 in) polishing pad. An industrial rated drill press capable of variable downforce (7-70 kPa +/- 1 kPa or 1-10 psi +/- 0.2 psi) via a weighted traverse replaces the standard RotoPol head. The traverse is mounted to the drill press in such a way so that the downforce applied is directly transmitted to the wafer and will not create a moment about the drill press itself. Pad conditioning can be done either in situ, during polishing, or ex situ, before and after polishing. We use a 50 mm diameter diamond grit conditioner wafer that both rotates and sweeps across the pad.

Slurry film thickness measurements are done using an optical technique known as dual emission laser-induced fluorescence or DELIF. This technique is described in detail in Coppeta and Rogers [13]. DELIF uses the fluorescence from two different commercially available dyes (mixed in with the slurry) each fluorescing at different wavelengths to measure slurry mixing, fluid depth, or fluid temperature. A complete description of these measurements can be found in our previous work [14-17]. The dyes are excited by a 100 Watt UV lamp source and we use two high-resolution 12-bit spatially aligned digital cameras to capture the fluorescence data beneath the wafer. The optics used on these cameras enable us to image an area of 2.25 cm by 3.8 cm

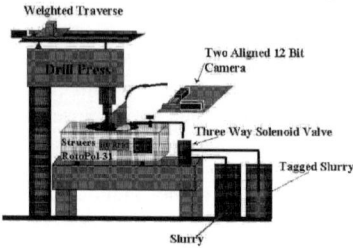

Figure 1: *Tabletop polisher and experimental Setup*

with a spatial resolution of approximately 50 μm per pixel. The images from each camera are aligned to within one pixel and can resolve fluid thickness variations down to 1 μm. Actual slurry thickness measurements are repeatable to within 5 μm. Since we are using optical techniques to measure the slurry flow beneath the wafer, actual silicon wafers cannot be used. Instead we use transparent BK-7 glass windows modified with gimbal mounts as our polishing substrates. The glass wafer is 75 mm (3 in) in diameter, and is typically bowed +/- 5 microns convex or concave. A convex wafer bows out, and a concave wafer bows in. (Figure 2 a&b)

We use a load cell mounted between two sliding plates below the polisher to measure friction during the polishing process. The bottom plate is fixed to a vibration isolation table and the polisher is fixed to the upper plate. The load cell will sense the friction force created by the interaction of the wafer and polishing pad to within 3.5% accuracy. Equation (1) shows the coefficient of friction , μ, as the friction force, F_F, normalized by the downforce, P.

$$\mu = \frac{F_F}{P} \qquad (1)$$

All polishing parameters and data acquisition including platen speed, downforce, conditioner speed, and slurry flow rate are computer controlled and monitored. The cameras are similarly computer controlled and can be synchronized with changes in polishing parameters so that we can acquire image data, as well as friction data, at any point during the polishing process. The polishing parameters that we will focus on in the following results are the effects polishing pad speed and wafer downforce. The conditioner arm sweeps across the pad at 10 oscillations per minute, and the wafer rotation rate is held constant at 60 RPM.

RESULTS

This study is conducted with both in situ and ex situ conditioning. The results presented here will only include that which is done with in situ conditioning. In situ conditioning provides a more consistent pad treatment than ex situ conditioning does where pad glazing can occur. Also, the slurry film thickness referred to here is a spatially averaged measurement. Since the pad topography largely dictates how much fluid it can trap between the asperities, the fluid depth is not exactly measured from the wafer surface to the pad surface, but rather an average depth of the pad asperities. For consistent pad topography, in situ conditioning maintains these asperities and pore sizes.

The curvature of the wafer or the degree to which the wafer is bowed has a significant effect on the fluid dynamics of the polishing process. Figures 3 and 4 each correlate the trends in slurry film thickness underneath the wafer to the frictional coefficient as the polishing pad speed is changed for a convex and concave wafer, respectively. From the plot in figure 3 the slurry thickness underneath a convex wafer increases, and the coefficient of friction decreases as the pad speed increases. At a low pad velocity of 30 RPM (relative pad-wafer velocity of 0.25 m/s) the wafer rides on about a 40 μm thick slurry layer. As the pad speed increases to 90 RPM (0.72 m/s) this slurry layer increases in thickness by about 10 μm. Also, the coefficient of friction decreases about 70% over this range. This suggests that the lubrication layer between the wafer

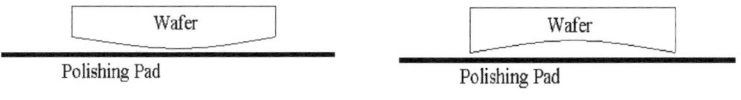

Figure 2a: *Convex wafer*	**Figure 2b:** *Concave wafer*

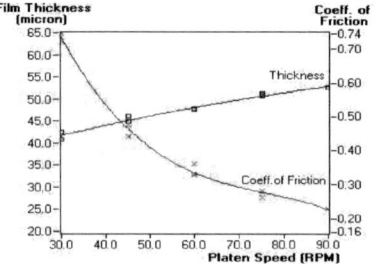

Figure 3: *Slurry film thickness and coefficient of friction as a function of pad speed for a convex wafer*

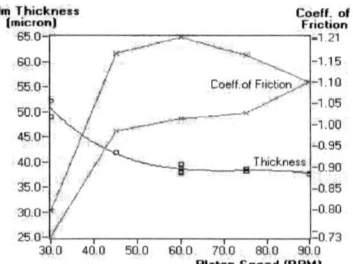

Figure 4: *Slurry film thickness and coefficient of friction as a function of pad speed for a concave wafer*

and the pad moves towards a full hydrodynamic lubrication regime at higher relative velocities, as seen previously by Coppeta et al [18]. At conditions of a full hydrodynamic lubrication regime the wafer is completely separated from the pad asperities by a fluid layer. This is in contrast to a partial or boundary lubrication regime where there is asperity contact with little or no fluid lubricating layer.

The concave wafer case is different from the previous convex wafer case. In figure 4, the slurry thickness decreases and the coefficient of friction increases as the pad speed increases. At a low pad velocity of 30 RPM the slurry film thickness is about 50 µm. As the pad speed increases to 90 RPM the slurry thickness decreases almost 15 µm. The coefficient of friction here increases 35% over the range of the same pad speeds. The trend with the concave wafer is clearly opposite from the convex. As the relative velocities increase, the concave wafer moves towards increased asperity contact meaning partial or total boundary lubrication. We have measured a net negative fluid pressure underneath the wafer for a concave wafer only. This suction brings the wafer and pad into contact, thereby, decreasing slurry thickness and increasing friction as seen by Tichy et al.

In order to isolate the effect of wafer downforce, the pad velocity is fixed at 60 RPM (relative wafer-pad velocity of 0.5 m/s) and the applied downforce is varied from 2 psi to 6 psi. In both cases of convex and concave wafers, the slurry film thickness decreases about 15 µm for a 4 psi increase in downforce. Although the trends are the same, the slurry layer thickness under the convex wafer is about 10 µm greater than that for the concave wafer. The manner in which

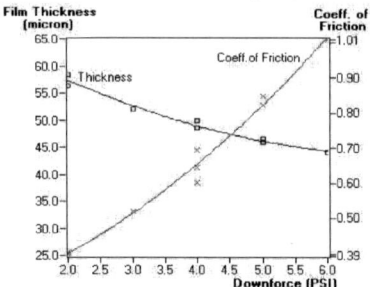

Figure 5: *Slurry film thickness and coefficient of friction as a function of downforce for a convex wafer*

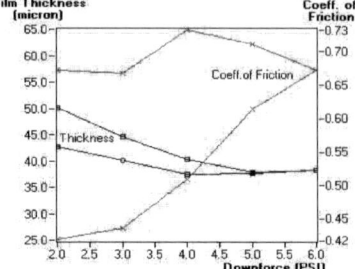

Figure 6: *Slurry film thickness and coefficient of friction as a function of downforce for a concave wafer*

the pad conforms to the wafer shape is believed to be one of the causes of this difference. The increased wafer downforce pushes the wafer into the compliant pad creating higher pad asperity pressure and decreasing relative fluid pressure. Therefore, less slurry resides in the gap between the wafer and the pad.

However, the behavior of the slurry film thickness and the coefficient of friction in the concave wafer case is not what we would expect. The expectation of the thickness and friction for the concave wafer in figure 6 would be the same trends as the convex wafer's in figure 5. From figures 4 and 6 the repeatability in the data of the concave wafer is compromised for two reasons. The wafer is polishing, therefore changing shape during this experiment, and the pad is deforming as well. In figure 6, as the downforce is decreased from 6 psi to 2 psi the slurry thickness increases to 50 μm, higher than it was originally (43 μm), at 2 psi. This experimental phenomenon has been repeated over a range of input parameters and at this time a possible explanation for this repeatable experimental result could be the effects of hysterisis in pad compliance. It will be necessary to understand the pad's response to changing downforce not only to characterize the lubrication regime, but also to understand how the pad preferentially polishes fine features on a patterned wafer.

The coefficient of friction, in figure 6, drops as the downforce reaches 6 psi and then continually decreases as the downforce is reduced back down to 2 psi. As the wafer goes from concave to increasingly convex the lubrication regime during polishing changes. As seen earlier, a convex wafer sees a far greater amount of hydrodynamic lubrication than a concave wafer. So as this experiment progresses the level of hydrodynamic lubrication increases, the frictional coefficient decreases, and the slurry thickness increases as the wafer hydroplanes. To truly isolate the effect of wafer shape the concavity of the wafer needs to be preserved.

Another way to look at the fluid dynamics between the wafer and polishing pad is to consider at the wafer angle of attack, or wafer tilt angle. We measure the fluid thickness under the front half and back half of the wafer and extrapolate the angle of attack from the difference in thickness. This technique can accurately measure angle of attack down to 0.003 degrees. Figure 7 is a comparison of this tilt angle as a function of wafer downforce for a convex and concave wafer. The angle of attack for a convex wafer is greater by orders of magnitude than that of a concave wafer. As the wafer downforce increases the convex wafer's angle of attack decreases, whereas the concave wafer's angle of attack relatively does not change. The substantially greater angle of attack of the convex wafer can continually support a larger hydrodynamic fluid layer. In the other case where the concave wafer's angle of attack is low, it becomes difficult for the wafer to maintain a hydrodynamic fluid layer.

CONCLUSION

The wafer curvature strongly affects the fluid behavior during CMP. We have examined

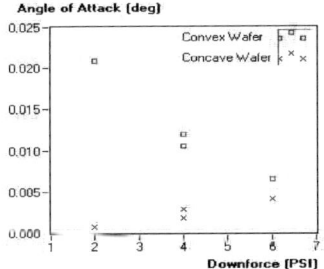

Figure 7: *Wafer angle of attack as a function of downforce*

both convex and concave wafers and their effects on slurry film thickness and the coefficient of friction. A convex wafer is found to support a hydrodynamic fluid layer much more easily than a concave wafer. As the relative pad-wafer velocity increases a convex wafer will ride on an increasingly thick slurry film, whereas a concave wafer will be sucked into the polishing pad increasing asperity contact and friction. As wafer downforce increases the slurry film thickness decreases. In this case the wafer shape determines the relative thickness of the slurry layer. The shape of the wafer has a large influence on the lubrication regime that exists between the wafer and polishing pad. In order to further understand the correlation between slurry film thickness and friction, additional knowledge of the polishing pad's response to pressure fluctuations is necessary. Through some accurate measurements of friction and slurry layer thickness during CMP, increased control of the process may lead to greater polishing performance.

ACKNOWLEDGEMENTS

The authors would like to thank Intel and Cabot corporations for funding this research. We would like to thank VEECO corporation for the donation of a Dektak 200 Si profilometer that enabled us to measure wafer topographies. We would also like to thank Freudenberg Nonwovens for donating FX-9 polishing pads.

REFERENCES

1. L. Cook, J. Non-Crystalline Solids **120**, 152-171 (1990).
2. Z. Stavreva, D. Zeidler, M. Plotner, K. Drescher, App. Surf. Sci. **108**, 39-44 (1997).
3. S. Runnels, J. Electrochem. Soc. **141**, 1900-1904 (1994).
4. J. Levert, R. Baker, F. Mess, R. Salant, S. Danyluk, STLE Trib. Trans., Submitted for publication Aug. 1997.
5. F. Mess, J. Levert, S. Danyluk, Wear **211**, 311-315 (1997).
6. J. Tichy, J. Levert, L. Shan, S. Danyluk, J. Electrochem. Soc. **146**, 1523-1528 (1999).
7. S. Sundararajan, D. Thakurta, D. Schwendeman, S. Murarka, W. Gill, J. Electrochem. Soc. **146**, 761-766 (1999).
8. J. Lu, J. Coppeta, C. Rogers, L. Racz, A. Philipossian, F. Kaufman, J. Electrochem. Soc. (submitted Nov.1999).
9. L. Cook, J. Wang, D. James, A. Sethuraman, Semiconductor Int'l. Nov. 1995, 141-144.
10. Y. Su, S. Wang, J. Hsiau, Wear **188**, 77-87 (1995).
11. M. Bhushan, R. Rouse, J. Lukens, J. Electrochem. Soc. **142**, 3845-3851 (1995).
12. S. Runnels, L. Eyman, J. Electrochem. Soc. **141**, 1698-1701 (1994).
13. J. Coppeta, C. Rogers, Experiments In Fluids 25, 1-15 (1998).
14. C. Rogers, J. Coppeta, L. Racz, A. Philipossian, F. Kaufman, D. Bramono, J. Elec. Mat. **27**, 1082-1087 (1998).
15. J. Coppeta, C. Rogers, L. Racz, A Philipossian, F. Kaufman, Proc. CMP-MIC Conf., Santa Clara, CA, 1999.
16. J. Coppeta, L. Racz, C. Rogers, A. Philipossian, F. Kaufman, Int'l J. of CMP for On-Chip Interconnection **1**, 47 (1999).
17. J. Coppeta, C. Rogers, L. Racz, A. Philipossian, F. Kaufman, J. Electochem. Soc. **147** (to be published May 2000).
18. J. Coppeta, J. Lu, D. Bramono, C. Rogers, L. Racz, A. Philipossian, F. Kaufman, 4th Int'l Symposium on CMP, Lake Placid, New York, Aug 8-11, 1999.

A MODEL OF CHEMICAL MECHANICAL POLISHING

Ed Paul Stockton College, Pomona NJ 08240 and NIST, Gaithersburg MD 20899 U.S.A.

ABSTRACT

A generic model is presented which explains the dependence of chemical mechanical polishing rates on the concentration of reacting chemicals and abrasives in the slurry. The predictions of this model are compared to data from the literature for tungsten CMP.

INTRODUCTION AND THEORY

CMP has been described qualitatively as an alternation of chemical reaction and mechanical abrasion processes[1]. This paper provides a generic, semiquantitative description for both of these processes and links these descriptions to form a model that predicts the removal rate as a function of the concentrations of chemicals and abrasives in the slurry. Certain modifications are needed to apply the generic model to specific systems. As an example, the results will be compared to polishing data available in the literature for tungsten.

In this model of CMP, the abrasive particles are small and the pad and slurry fluid support the load. Chemical reactions between the workpiece material and chemicals in the slurry form a thin film on the workpiece surface. This film, which is not tightly bonded to the bulk material, is separated from the workpiece surface by abrasive particles, which are pushed into it by the polishing pad. Fresh surface is then exposed on the workpiece, and this surface is available for reaction with the chemicals in the slurry for a subsequent CMP cycle. The chemical process of forming the reaction surface film depends on how much of the reacting chemical is in the slurry. At low chemical concentrations, the surface is only partially covered. Increases in chemical concentration will increase the surface coverage until the film is complete and additional chemical has no effect on the polishing rate. Similarly, the mechanical process of removing this film depends on the amount of abrasive in the slurry. At high abrasive loading, the pad is full and adding more abrasive will not affect the polishing rate, while at low abrasive loading there is space on the pad surface to hold more abrasive particles. In this case, additional abrasive will increase the polishing rate. A quantitative discussion of these processes is presented below.

Chemical Reaction Process In the chemical process, the workpiece material reacts with chemical components of the slurry to form a thin reaction film. This surface reaction is subject to the laws of chemical equilibrium, and can be written schematically as

$$M + C \leftrightarrow MC^* \qquad \text{Rxn. 1}$$

where M and C represent the workpiece material M and the reacting chemical C while MC* represents the surface complex which forms the film. The reaction is a reversible one: the film can decompose and return chemical C to the slurry. The MC* complex is available for either dissolution into the slurry or for mechanical abrasion. When stable surface films form, their dissolution, sometimes called corrosion, is a slow process. It may be written as

$$MC^* \rightarrow MC_{(aq)} + M \qquad \text{(Rxn. D)}$$

where MC$_{(aq)}$ represents the product dissolved into the slurry and M is the workpiece surface material, revealed when the surface complex is removed. The abrasion step may be written as

$$MC^* + A \rightarrow MC\text{-}A + M \qquad \text{(Rxn. M)}$$

for abrasive particles A, with material MC-A leaving the workpiece surface to expose fresh M.

Chemical kinetics predicts rates using rate constants for simple reactions. In this case, reversible reaction 1 has associated rates for the forward (f) and reverse (r) reactions

$$r_{1f} = k_{1f}\, N_M\, [C] \qquad \text{(1f)}$$
$$r_{1r} = k_{1r}\, N_{MC^*} \qquad \text{(1r)}$$

where k_i is the rate constant for reaction i, N_M and N_{MC^*} are the number of M and MC* sites on the workpiece surface, and $[C]$ is the concentration of reacting chemical in the slurry. The rate for dissolution is proportional to N_{MC^*} while the rate for mechanical abrasion depends on both N_{MC^*} and the number of effective abrasive particles N_A per area A.

$$r_D = k_D\, N_{MC^*} \qquad \text{(2)}$$
$$r_M = k_M\, (N_A/A)\, N_{MC^*} \qquad \text{(3)}$$

The total removal rate is the sum of these corrosion and abrasion rates, and is proportional to the number of surface complexes N_{MC^*}. Standard methods of chemical kinetics for surface reactions show that the rate of change of N_{MC^*} is

$$d\, N_{MC^*}/dt \;=\; k_{1f}\, N_M\, [C] \; - \; k_{1r}\, N_{MC^*} \; - \; k_D\, N_{MC^*} \; - \; k_M\, (N_A/A)\, N_{MC^*} \qquad \text{(4)}$$

which equals 0 at steady state, when the rate of formation of MC* balances the rate of removal. The total number of surface sites N_{oM} is related to the workpiece area A by the area per site, $d_M{}^2$.

$$N_{oM} \;=\; N_M + N_{MC^*} = A/d_M{}^2 \qquad \text{(5)}$$

At steady state, combining Eqs. 4 and 5 gives

$$N_{MC^*} \;=\; (N_{oM})\,[C]\,/\,(\,[C] + (k_{1r} + k_D + k_M(N_A/A)\,)/\,k_{1f}\,) = (A/d_M{}^2)\,\theta \qquad \text{(6)}$$

where $\theta = N_{MC^*}/N_{oM}$ is the fraction of surface sites covered by the MC* complex. This expression will be used to find the polishing rate in Section IV.

Mechanical Abrasion Process The mechanical abrasion rate depends on the number of effective abrasive particles in the slurry that are available for contact with the workpiece. In this model, the polishing pad has a limited total number of surface sites, N_{oP}, which hold abrasive particles. At any given concentration of abrasive, some of these sites will be occupied and others will be empty. The fraction of occupied sites depends on the abrasive concentration: Let r_{Af} be the rate at which an abrasive particle enters a pad site from the slurry. This is proportional to the number of available sites N_s, the concentration of abrasives in the slurry $[A]$, and the rate constant k_{Af} giving $r_{Af} = k_{Af}\, N_s\, [A]$. The rate at which abrasives leave the pad is proportional to the number of occupied sites N_A, so $r_{Ar} = k_{Ar}\, N_A$. The total number of sites on a pad may be written as $N_{oP} = N_s + N_A$. At steady state, the forward and reverse rates balance. Combining these equations gives N_A, the number of effective abrasives on pad area A.

$$N_A = N_{oP}\,[A]\,/\,(\,[A] + (k_{Ar}/k_{Af})\,) = A\,c_P\,[A]\,/\,(\,[A] + K_A\,) = A\,c_P\,\theta_A \qquad \text{(7)}$$

where $c_P = N_{oP}/A$ is the site density on the pad, $K_A = (k_{Ar}/k_{Af})$ is the pad-abrasive equilibrium constant, and θ_A is the fraction of pad sites occupied by abrasive particles.

The rate of abrasion is proportional to the surface density of available abrasives N_A/A and to the number of MC* surface complexes on the workpiece N_{MC*}. The proportionality constant k_M is a function of the pad pressure and velocity, the abrasive diameter d_A, and the mechanical and fluid dynamic properties of the pad and slurry.

$$r_M = k_M (N_A/A) N_{MC*} = k_M (c_P \theta_A) \theta (A/d_M^2) \qquad (8)$$

Overall CMP Rate Combining equations 2, 6 and 8 gives the polishing rate (material removal per workpiece area) R as a sum of dissolution and mechanical abrasion processes.

$$R = (r_D + r_M)/A = (k_D + k_M c_P\theta_A) \theta / d_M^2 \qquad (9)$$

In this expression, the dependence of the polishing rate on the concentration of chemical in the slurry [C] is contained in the surface coverage parameter θ, while the dependence on particle concentration [A] is contained in the pad coverage fraction θ_A. Converting [A] to the experimentally accessible weight percentage %A and combining the various chemical and mechanical parameters into the constants β_i, Eq. 9 can be transformed to show explicitly the dependence on [C] or %A as

$$R = \beta_1 [C] / ([C] + \beta_2) \qquad (10)$$
$$R = \beta_3 + \beta_4 \%A / (\%A + \beta_5) \qquad (11)$$

Both of these equations show an initial increase in polishing rate (with slope β_1/β_2 for [C] and β_4/β_5 for %A) as function of concentration, and asymptotic polishing rate limits (β_1 and $\beta_3 + \beta_4$) at high concentration. The removal rate approaches 0 as [C] approaches 0, and approaches β_3 as abrasive loading approaches 0. This behavior is in agreement with experimental results in a variety of systems. Explicit results for tungsten are shown in the next section.

RESULTS : TUNGSTEN CMP

Experimental results for the dependence of polishing rate on chemical and abrasive concentrations for W-CMP have been published by a number of authors[2-4]. The chemical process for tungsten is somewhat more complex than the generic model. The surface film formed by the reaction between W and the chemical oxidizers in the slurry has been shown[5] to develop in two distinct phases. At first an oxide monolayer forms slowly, followed by a faster reaction which leads to a limiting film as perhaps 9 additional layers of oxide are added through an anion vacancy diffusion mechanism to form a mature WO_3 surface. This kinetic film may not be in thermodynamic equilibrium and the oxide layers may be non-stoichiometric, varying between WO_2 and WO_3. The observed[3] independence of surface roughness from abrasive size is due to the delamination of the mature film. In modeling terms, reaction 1 must be replaced by a sequence of reactions for the formation of the initial oxide WOx and the mature oxide WOxn

$$W + Ox \leftrightarrow WOx \rightarrow WOxn \qquad (Rxns\ W1 , W2)$$

with rate expressions

$$r_{1fW} = k_{1f} N_W [Ox] \qquad r_{1rW} = k_{1r} N_{WOx*} \qquad r_2 = k_2 N_{WOx*} \qquad (W1f,r , W2)$$
$$r_D = k_D N_{WOxn*} \qquad r_M = k_M (c_P \theta_A) N_{WOxn*} \qquad (W_D , W_M)$$

where the formation of the mature WOxn* surface complex is needed for material removal by dissolution or mechanical abrasion. It is helpful to define purely chemical terms for this system:

$$K_1 = k_{1f} / (k_{1r} + k_2) \qquad\qquad F_C = k_2\, K_1[Ox] / (1 + K_1[Ox]) \qquad (K_1, F_C)$$

Following the same methods as in the generic model above leads to equations 10 and 11, with explicit predictions for the parameters β_i as shown in Table 1. If the rate at which the surface film dissolves is small compared to the abrasion rate, then k_D can be set to 0, giving simpler expressions for β_i in general, and β_3 in particular. It is helpful to note that all of the β_i have the form of a product of chemical and mechanical terms divided by their sum. This form forces a balance between the chemical and mechanical processes: if either becomes much larger than the other, then its effect will be negligible.

Table 1. Expressions for β_i

	General Formula	$k_D = 0$
β_1	$\dfrac{k_2\left(k_D + k_M c_P \theta_A\right)}{d_W{}^2\left(k_2 + k_D + k_M c_P \theta_A\right)}$	$\dfrac{k_2\left(k_M c_P \theta_A\right)}{d_W{}^2\left(k_2 + k_M c_P \theta_A\right)}$
β_2	$\dfrac{k_D + k_M c_P \theta_A}{K_1\left(k_2 + k_D + k_M c_P \theta_A\right)}$	$\dfrac{k_M c_P \theta_A}{K_1\left(k_2 + k_M c_P \theta_A\right)}$
β_3	$\dfrac{k_D F_C}{d_W{}^2\left(k_D + F_C\right)}$	0
β_4	$\dfrac{F_C}{\left(k_D + F_C\right)}\dfrac{F_C\left(k_M c_P\right)}{d_W{}^2\left(k_D + F_C + k_M c_P\right)}$	$\dfrac{F_C\left(k_M c_P\right)}{d_W{}^2\left(F_C + k_M c_P\right)}$
β_5	$\dfrac{K_P\left(k_D + F_C\right)}{\left(k_D + F_C + k_M c_P\right)}$	$\dfrac{K_P F_C}{\left(F_C + k_M c_P\right)}$

Eqs. 10 and 11 have been used to fit four different sets of tungsten polishing rate data from the literature. In each of these cases, the model fits the data well. Fig. 1 shows the polishing rate as a function of chemical concentration for three different pad pressure and velocity conditions, with rate increases at low concentration approaching a plateau at high concentrations[2]. The fitting parameters β_1 and β_2 vary monotonically with pressure and velocity.

Fig. 2 shows similar data as a function of abrasive concentration, with rate increases at low concentration approaching a plateau at high concentrations[2]. In this case data fitting gives $\beta_3 = 0$ implying that the dissolution rate is much slower than abrasion. β_4 and β_5 vary monotonically with pressure and velocity.

Fig. 3 shows the polishing rate as a function of abrasive loading[4] for 5 different sizes of Al_2O_3 abrasive. The data fitting gives $\beta_3 = 0$ (dissolution rate slower than abrasion) while β_4 and β_5 vary with abrasive size. The dependence of rate on abrasive size is partially correlated with the particle concentration: for a given weight % abrasive, there are more small particles than large ones, and thus there are more abrasive contacts. As a function of particle concentration in particles/cm^3 all of the points in Fig. 3 can be put onto one curve. Because polishing conditions are different however, the β_i values do not compare directly with those from Figs. 1 and 2.

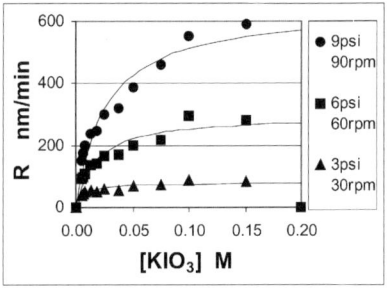

Fig. 1. Polishing rate vs KIO$_3$ concentration at different P and v conditions. Data as given in Ref. 2, curves from Eq. 10.

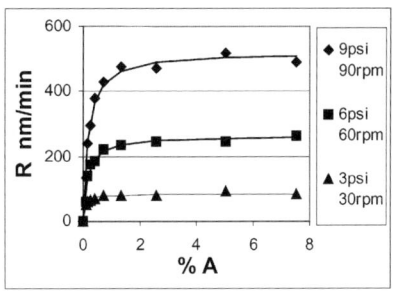

Fig. 2. Polishing rate vs abrasive loading at different P and v conditions. Data as given in Ref. 2, curves from Eq. 11.

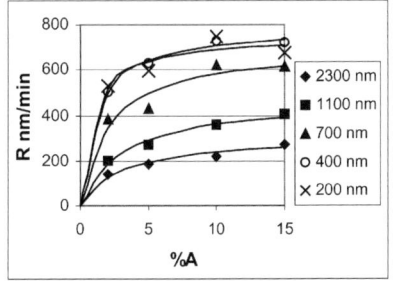

Fig. 3. Polishing rate vs abrasive loading for different abrasive diameters. Data from Ref. 3, curves from Eq. 11.

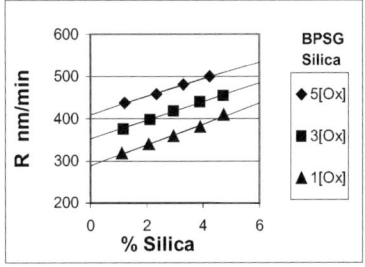

Fig. 4. Polishing rate vs abrasive loading at different relative concentrations of oxidizer for BPSG patterning and silica abrasive. Data from Ref. 4, curves from Eq. 11

Fig. 4 compares experimental data with the predictions of Eq. 11 for the polishing rate of tungsten in patterned wafers at different relative concentrations of oxidizer and at different abrasive loading[4]. The identity and actual oxidizer concentrations were not given, but the increase of polishing rate with oxidizer concentration is qualitatively predicted by Eq. 10. Fig. 4 is one of three similar series of data in which different abrasives (silica and alumina) and different interlayer dielectric patterning materials (BPSG and TEOS) were used. The results of fitting all three sets of data to Eq. 11 showed that, in contrast to results with unpatterned wafers, β_3 is not 0, indicating that the dissolution rate is large compared to abrasion. β_3 does depend on the concentration of oxidizer, in agreement with Eq. 10. The values of β_3 for both alumina and silica abrasives were the same for BPSG patterning, while β_3 for BPSG differed from that for TEOS when alumina was used as the abrasive. This shows that the rate of abrasion for patterned wafers is dominated by the dielectric used while the dissolution rate is independent of abrasive.

β_4 values depend on the concentration of oxidizer, but are independent of the abrasive and dielectric at low [Ox]. At high [Ox] the TEOS patterning with less dissolution shows faster abrasion than the BPSG system. At lower [Ox] dissolution dominates and the growth of the surface film is apparently too small to differentiate these systems. β_5 values are different for the different systems and conditions.

DISCUSSION

Eqs. 10 and 11 are simply 2 parameter data fits (when dissolution is negligible) which are approximately linear over certain regions. However, their basis in a process model focuses attention on specific process questions. The overall view is that the polishing rate depends on both mechanical r_M and chemical r_C parameters as in Eq. 12.

$$R = r_M r_C / (r_M + r_C)$$ (12)

This expression implies that the polishing rate can be increased by increasing either the mechanical or chemical rates separately, but only up to some limiting value. The limiting value itself can be increased by simultaneously changing both chemical and mechanical parameters.

The mechanical parameters include the process conditions of pressure, velocity and abrasive concentration, any of which can be varied continuously. There is some evidence that different abrasives will give different polishing rates and that abrasive selection could also affect polishing. Slurry viscosity and pad composition also play roles in the mechanical removal rate.

Chemical parameters include concentration of reactant, which can be varied continuously, and choice of reactant. Different chemicals will react faster or slower with specific surfaces. Choosing the appropriate slurry chemistry would permit shifts from one set of limiting rates to another. pH and temperature also affect chemical reaction rates. This model suggests that it is possible to use chemical reaction rates, obtained in simpler slurry systems without abrasives present, to predict polishing removal rates. For example, different oxidizers - KIO_3, $K_3Fe(CN)_6$, H_2O_2, and $Fe(NO_3)_3$ - have all been used successfully in tungsten CMP. Each has different reaction rate constants and thus different values for K_1 and F_C in Table 1.

It is possible to empirically adjust either pressure or chemical activity to get a practical polishing rate. However, a model can guide this search by predicting directions for useful experiments and by helping to understand the results obtained.

ACKNOWLEDGEMENTS

The author is grateful to Chris Evans for his suggestions and for the opportunity to be a guest worker at NIST, and to Professor S. V. Babu for his helpful comments.

REFERENCES

1. Kaufman, F. B.; Thompson, D.B. et al. *J. Electrochem. Soc.* **138**, 3460-3465 (1991)
2. Stein, D. J.; Hetherington, D. L.; Cecchi, J. L. *J. Electrochem. Soc.* **146**, 376-381 (1999)
3. Bielmann, M.; Mahajan, U.; Singh, R. K. *Electrochem. Solid-State Lett.* **2**, 401-403 (1999)
4. Jairath, R., Desai, M., Stell, M., and Tolles, R. *Mater.Res.Soc.Proc.* **337**, 121-131 (1994)
5. Fauconnier, J.; Vennereau, P. *Electrochim. Acta* **23**, 113-119 (1978)

Mat. Res. Soc. Symp. Vol. 613 © 2000 Materials Research Society

Modeling on Mechanical Properties of Polishing Pads in CMP Process

Takeshi Nishioka*, Satoko Iwami*, Takashi Kawakami*,
Yoshikuni Tateyama**, Hiroshi Ohtani** and Naoto Miyashita**
* Mechanical Systems Laboratory, Corporate R & D Center, Toshiba Corporation
1, Komukai-Toshiba-cho, Saiwai-ku, Kawasaki, 212-8582, JAPAN
** Semiconductor Company, Toshiba Corporation
8, Sugita-cho, Isogo-ku, Yokohama, 235-8522, JAPAN

ABSTRACT

Chemical mechanical polishing is an essential process for achieving a high degree of planarization. The planarity after CMP sensitively depends on pattern scales, pattern densities and mechanical properties of polishing pads. In order to simulate the topography after CMP, a numerical model for the polishing pad is proposed. In this model, the surface roughness layer of the polishing pad is assumed as a flat soft layer. The distribution of the contact pressure between the patterned wafer and the polishing pad is calculated with finite element method, and the pattern topography is modified based on the pressure dependency of the polishing rate. The iterations of the contact pressure analyses and the topography modifications give the progress of the polishing process numerically. The model is applied to oxide CMP process with silica slurry and stacked pad of polyurethane and non-woven fabric. The compressive elastic moduli of polyurethane layer and non-woven fabric layer are measured dynamically. The elastic modulus of the soft layer is treated as a fitting parameter between the experimental results and the numerical model. The models with the elastic modulus of 10 MPa for the soft layer show good agreements with the experimental results in both of a short range, where the compressive deformation of the pad is dominant, and a long range, where the bending deformation is dominant. Static measurements for the surface elasticity of the polyurethane layer also give a good agreement with the model. The proposed pad model should be useful for the topography simulation, and it also guides the development of new polishing pads.

INTRODUCTION

CMP is now recognized as the planarization method of pre-metal and inter-layer dielectric and metal via and damascene. It achieves good global planarity over chip and wafer, however, the requirements of planarity and uniformity progressively become severe for future CMP processes. Planarity after CMP largely depends on the pattern scales, the pattern densities and the mechanical properties of the polishing pads. Therefore the contact pressure distribution between the wafer and the pad, including contact through the abrasives, and the pad deformations are very important for understanding the planarization process.

Several models were proposed for CMP and reviewed [1]. Some models include the effects of the pad bending deformations on the wide range planarity. Ohtani, one of the authors, reported a spring model for the compressive deformations as a simple and accurate evaluation [2]. The bending deformation is dominant for the global planarity and the wafer scale uniformity, and the compressive deformation is critical for the local planarity, i.e. dishing and thinning.   In this paper, a new modeling method on the mechanical properties of the polishing pad, which is effective for both of bending and compressive deformation, is proposed, and evaluated for the conventional oxide CMP process with silica slurry and polyurethane pad with non-woven fabric under layer.

MODELING

Figure1 shows a schematic of the contact between the patterned wafer and the polishing pad during CMP process. The deformation of the pad causes so-called dishing. The step height before CMP is less than 1 micrometer for dielectric layer. The surface roughness of the pad, which is not expressed in figure1, is much larger than the step height. Figure 2 shows the cross-sectional view of the conventional polyurethane pad. The pad contains a lot of bubbles of tens of micrometers. These bubbles form the roughness of 40 to 60 micrometers at the surface, and reserve slurry during the polishing process. The pad surface roughness is very important to discuss the contact behavior between the wafer, slurry and the polishing pad.

Previously, the authors discussed the hydrodynamic effects of the pad surface roughness using three-dimensional sinusoidal surface model, and concluded the effects was very small for the conventional oxide CMP process [3]. Therefore the down force applied to the wafer is supported by solid contacts between the wafer, the abrasives and polishing pad. The size of the abrasives is small enough compared with the step height and the pad surface roughness, to neglect the effects of the abrasives in the estimation of the polishing pad deformation. Finite element method (FEM) is effective for the contact pressure analysis between the wafer and the pad [4]. To include the effects of the topography change, the analysis flow shown in figure 3 was applied.

The surface roughness layer should be more easily deformed, compared with the bulk layer. Therefore the roughness layer was modeled as a flat "soft" layer. Figure 4 shows the FEM model for the contact pressure analyses. The polyurethane pad and the non-woven fabric under layer are stressed dynamically, because of the relative movement of the patters on the wafer and the pad rotation itself. The elastic moduli of the polyurethane bulk layer and non-woven fabric layer were measured dynamically by the configuration shown in figure 5. In order to obtain enough displacements and strain rate, the module of the polyurethane pad was measured with pressure

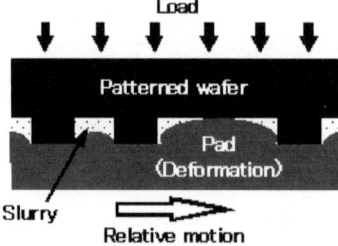

Figure 1. *Schematic of contact between patterned wafer and pad*

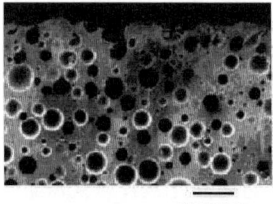

Figure 2. *Cross-sectional view of the polyurethane pad*

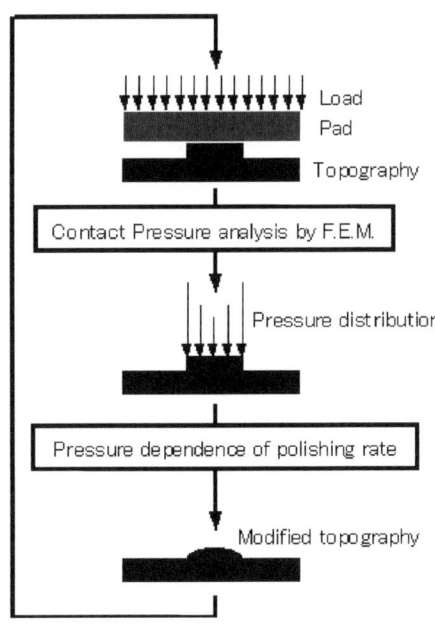

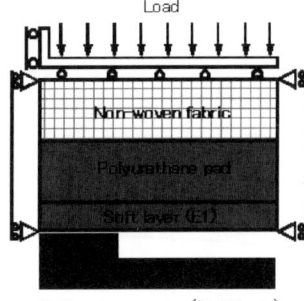

Pattern topography (Rigid body)

Figure 4. *FEM model*

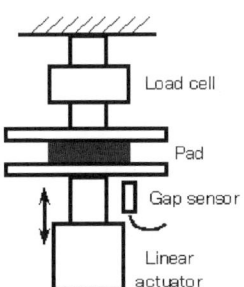

Figure 3. *Analysis flow*

Figure 5. *Configuration of measurement of the elastic module*

of 0.5 – 5.5 MPa and frequency of 100 Hz, and the module was 290 MPa. For the non-woven fabric layer, the module was 5.9 MPa measured with pressure range of 5 – 55 KPa and frequency of 1 Hz. The thickness of the soft layer was determined as 30 micrometers. The elastic module of the soft layer (E1) was treated as a fitting parameter. The wafer was assumed to be rigid and only the deformation of the pad was analyzed.

The pressure dependence of the polishing rate was investigated using thermal oxide blanket wafers. The results were shown in figure 6. The relationship could be expressed as Preston's linear equation [5]. In order to determine the elastic module of the soft layer, the planarization process of 30-micrometer wide line and 300-micrometer wide space pattern was investigated by experiments and analyses. Mean pressure (Down force) was 30 KPa. Analyses were attempted with 50, 10 and 5 MPa for E1 and the time increment was 1 second, i.e. 60 time iterations of FEM and topography modification for 60 second polishing.

Comparison between the results of experiments and analyses were shown in figure 7. The vertical axis presents the remaining step height after polishing and the horizontal axis shows the removal at the center of 30-micrometer wide line. The analysis with 50 MPa for E1 showed the ideal planalization, i.e. no dishing. The model with the module of 10 MPa had very good

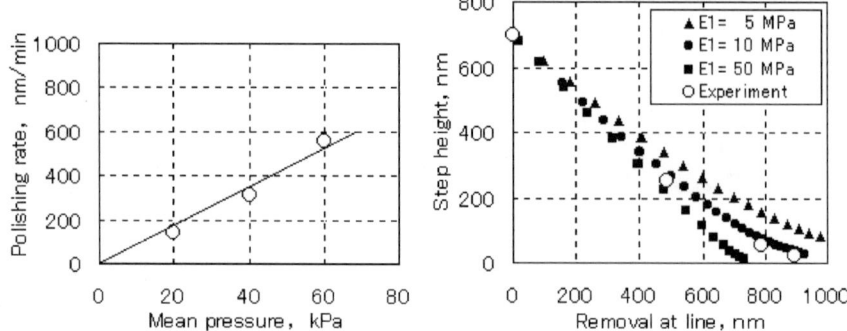

Figure 6. *Pressure dependence of polishing rate*

Figure 7. *Comparison between experiments and analyses*

agreements with the experimental results. Based on the parameter survey, the elastic module of the soft layer was fixed as 10 MPa.

EVALUATION AND DISCUSSION

The pad model with the surface soft layer of 30 micrometer thickness and 10 MPa elastic module was applied to other topography. The right of figure 8 shows the results for 300 micrometer wide line and 300 micrometer wide space pattern. The analytical results had good agreements with experimental ones. Because the thickness of the polyurethane pad was 1.2 mm, the compressive deformation should be dominant for these planarization processes shown in figure 8. One the other hand, the bending deformation becomes effective for the planarization of large area, of which size is comparable to the pad thickness. Figure 9 shows the model evaluation for the multiple line and space pattern. The model had good agreements with the experiments even in this scale.

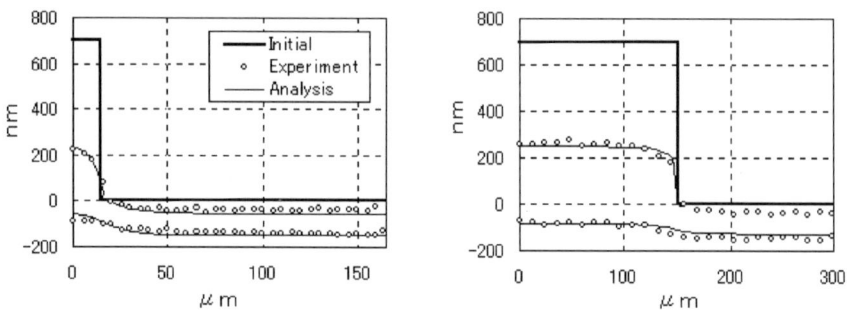

Figure 8. *Evaluation of the model for cell scale*
(Left : Line / Space = 30 um / 300 um , Right : 300 um / 300 um)

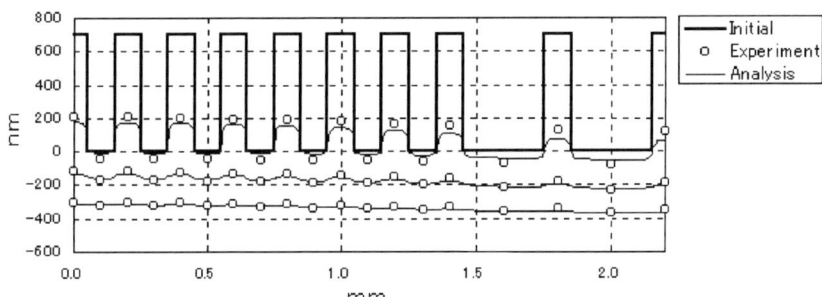

Figure 9. Evaluation of the model for cell scale
(Line / Space = 100 um /100 um , 100 um / 300 um)

Using axial symmetric elements for FEM, the model could be expanded into the wafer scale analysis. Because of the stress concentration and the stacked structure of the polishing pad, the uniformity of the removal is disturbed at the wafer edge in conventional oxide CMP. This phenomenon is regarded as the effect of the bending deformation of the pad. Figure 10 shows the results of the expanded analysis and experiment. 8 Inch wafer was used and the mean pressure was 50 KPa. The removal decreases at 97 mm point shows a fairly good agreement for the experiment and the analysis. It is pointed out that the model conforms to the wafer scale pad deformation. The discrepancy of the removal at 95 mm point is regarded as the effects of the backing film and the wafer deformation, which is not included in the present contact model.

In order to examine the soft layer, the surface elasticity of the polyurethane pad were measured. A cylindrical probe of 0.5 mm diameter was pressed to the pad surface. The load was 3 mN and the displacement was measured. More than ten data were obtained and the elastic module was calculated. The module ranged from 1.0 to 3.9 MPa, and average was 2.15 MPa. The measurements prove the soft contact behavior of the pad surface and the value of the elastic module is roughly close to that of the proposed model. Farther investigation is necessary for the discrepancy between the measured surface elastic module and proposed one.

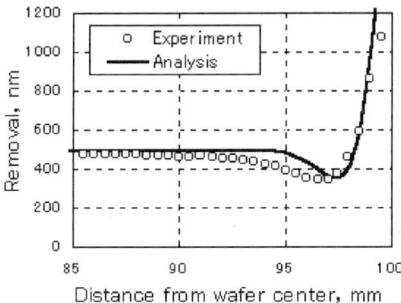

Figure 10. Evaluation of the model for
wafer scale

As mentioned before, the surface roughness layer is considered as reservoir for slurry during polishing. Its soft contact behavior should be also essential to avoid the stress concentration by large abrasives, which might cause the critical scratches. The model with the surface soft layer presents a good physical model not only for the planarity and the uniformity of the CMP process, but also for the scratches that we should avoid in the manufacturing.

CONCLUSIONS

Pad surface roughness layer, which is very important for understanding the contact behavior between the wafer and the pad, was assumed to be a flat and soft layer in the presented model for the conventional oxide CMP process with the silica slurry and the polyurethane pad. The analyses, using the contact pressure analyses by FEM and the topography modifications based on the pressure dependency of the removal rate, showed good agreements with the experiments not only in cell-scale but also in wafer-scale. The model accurately described both of the compressive and bending deformations. It presents a good physical model for polishing pads and should be useful for understanding planarization mechanisms and mechanical properties of polishing pads in CMP process.

ACKNOWLEDGMENT

The authors would like to send acknowledgment to Mr. Kinoshita and Dr. Shigyo in Semiconductor company of Toshiba for their encouragement and supports. The authors also thank Mr. Yano, Mr. Yoshii and Mr. Shoji in Toshiba for their fruitful discussions and technical supports.

REFERENCES

1.	G. Nanz and L. E. Camilletti, *IEEE Trans. Semiconductor Manufacturing*, **8**, 382 (1995)
2.	H. Ohtani, M. Murota, M. Norishma, H. Shibata and M. Kakuma, *Proc. 12th VLSI Multilevel Interconnection Conf.* 445 (1995)
3.	T. Nishioka, K. Sekine and Y. Tateyama, *Proc. IEEE 1999 Int. Interconnect Tech. Conf.* 89 (1999)
4.	H. Ohtani, Y. Tateyama, T. Nishioka, T. Matsuno and H. Shibata, *43rd Spring Meeting Extended Abst. of JSAP*, **2**, 752 (1996)
5.	L. M. Cook, *J. Non-Crystalline Solids*, **120**, 152 (1990)

Mat. Res. Soc. Symp. Vol. 613 © 2000 Materials Research Society

Fundamental Studies on the Mechanisms of Oxide CMP

Uday Mahajan, Seung-Mahn Lee and Rajiv K. Singh
Department of Materials Science and Engineering and Engineering Research Center for Particle Science and Technology, University of Florida, Gainesville, FL 32611

ABSTRACT

In this paper, results of studies on the addition of salt to a polishing slurry, in terms of its effect on slurry stability, SiO_2 polishing rate and surface roughness of the polished surface are presented. Three salts, viz. LiCl, NaCl and KCl were selected, and three concentrations were tested. Polishing rate measurements using these slurries show that adding salt leads to increased removal rate without affecting surface roughness significantly. Based on these results, we can say that the agglomerates formed by adding salt to the slurry are fairly soft and easily broken during the polishing process. In addition, turbidity and particle size measurements show that significant coagulation of the particles in the slurry occurs only at the highest salt concentration, and is fastest for LiCl and NaCl, with KCl showing the slowest coagulation. From these results, it can be concluded that the enhancement in polish rate is due to increased contact at the wafer-pad-slurry interface, and not due to formation of larger agglomerated particles in the slurry. This is because of reduced electrostatic repulsion between these three surfaces, due to the screening of their negative surface charge by the metal ions in solution, resulting in a higher wear rate.

INTRODUCTION

With the advent of multilevel metallization, Chemical Mechanical Planarization (CMP) has become one of the fastest developing areas in the microelectronics industry [1]. However, most of the knowledge in this field is based on previous studies on glass polishing [2]. In addition, a lot of CMP processes have been designed empirically, with little scientific understanding of the underlying phenomena. The effect of salt addition on polishing properties of slurries in particular is yet to be fully understood. Several salts are used in preparing slurries for CMP applications. They could be oxidizers, buffers or complexing agents. Addition of these salts is known to have an adverse effect on the stability of the slurry. This is chiefly due to the screening of the surface charge on the abrasive particles, which causes a compression of the electrical double layer around them. This results in a reduced repulsion between the particles, causing them to aggregate and settle in the slurry [3]. This settling is an undesirable phenomenon, as over a period of time these aggregates become hard to redisperse. These hard agglomerates can often damage the wafer surface, causing scratches and thereby reducing process yield. Although agglomeration of particles in high ionic strength environments has been extensively studied, a systematic investigation and correlation to polishing characteristics has not yet been carried out. In this study, we have reported the effect of addition of alkali metal salts (chlorides) on the polishing characteristics of SiO_2 slurries.

EXPERIMENT

The abrasives used for this study were 0.2 μm spherical sol-gel silica particles obtained from Geltech Corporation. These particles are spherical and have a very narrow size distribution, thus making them ideal as model particles for experimental studies. These particles (5.0 % by weight) were dispersed in DI water using an ultrasonic probe, following which the salt was added. ACS grade LiCl, NaCl and KCl were used, and the salt concentrations tested were 0.01 M, 0.1 M and 1.0 M. After the salt was added and dissolved, the slurry pH was adjusted to 10.50. The slurries were kept stirred continuously to minimize settling of particles. The samples used for polishing were 2.0 μm thick PECVD SiO_2 films deposited on p-type <100> silicon. A Struers Rotopol 31 polisher was used to conduct the polishing experiments, along with IC 1000/Suba IV stacked polishing pads (from Rodel Inc.). The polishing conditions used were as follows: a platen and head speed of 150 rpm (an average linear speed of 275 ft./min.), a down force of 7.0 PSI and a slurry feed rate of 100 ml/min. Three polishing runs were carried out for each slurry, and the polishing pad was conditioned at the beginning and between runs. A J.A. Woollam variable angle spectroscopic ellipsometer was used to measure film thicknesses before and after polishing, from which removal rates were determined. The surface morphology of the polished samples was characterized using a Digital Instruments Nanoscope III Atomic Force Microscope (AFM). In addition, settling (slurry turbidity) tests, particle size measurements and zeta potential studies were also conducted on the slurries as a function of ionic strength. An Acoustosizer, manufactured by Colloidal Dynamics was used for conducting the zeta potential measurements. This instrument utilizes the electroacoustic effect to determine particle size and charge (Zeta Potential), and can be used to characterize concentrated dispersions [4]. A Hach 2100 AN turbidimeter was used for conducting the turbidity measurements. The change in turbidity of the suspensions with time was used as an index of their stability. Particle size analysis was carried out using a Coulter LS 230 instrument, which uses dynamic laser scattering to obtain particle size.

RESULTS AND DISCUSSION

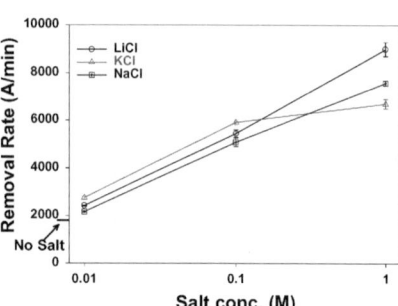

Fig. 1 shows polish rate as a function of salt concentration for the three different salts. It can clearly be seen that addition of salt causes a significant enhancement of polish rates, compared with the polish rate without salt addition, which is marked on the y-axis. This is expected to be due to formation of large aggregates in the slurry, which are formed as a result of the reduced repulsion between the abrasive particles in the presence of salt, causing them to come together. These larger particles should result in increased roughness of the polished surface.

Figure 1: Oxide Polish Rate as a function of salt concentration

However, from Fig. 2, it can be seen that this is not the case. Even under conditions of 1.0 M salt, the roughness of the polished surface is comparable to that of the surface polished by slurry

containing no salt. Similar results have been reported by us earlier for tungsten CMP [5] [6], where we have shown that polishing with slurries containing 0.1 M oxidizer (potassium ferricyanide) results in very smooth surfaces, although $K_3Fe(CN)_6$ is a multivalent salt and coagulates the alumina slurry in very small amounts. This implies that the shear stresses imposed by the polishing process are sufficient

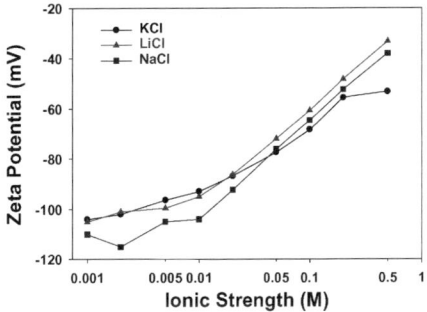

Figure 2: Surface Roughness of polished samples for different salt concentrations

to break any aggregates formed by adding salt, and that these larger particles are not responsible for the higher wear rates. Zeta potential results shown in Fig. 3 show that addition of salt suppresses the electric double layer around the silica particles and reduces their surface charge, making them more likely to agglomerate. The figure only shows the results of measurements carried out till salt concentrations of 0.5 M, as further additions of the salt increased the measurement error beyond acceptable limits. However, at 0.5 M salt it can be seen that the reduction in zeta potential is greater for LiCl and NaCl as compared to KCl. This suggests that these two salts are more effective at coagulating the slurry than KCl. These results were confirmed by turbidity measurements, which show no

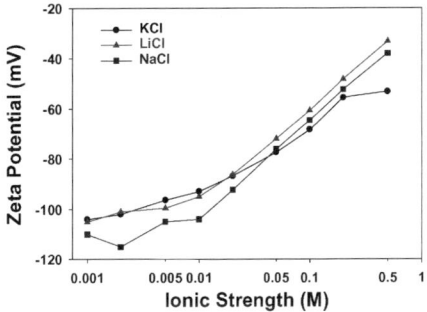

Figure 3: Zeta potential of SiO_2 slurries as a function of salt concentration

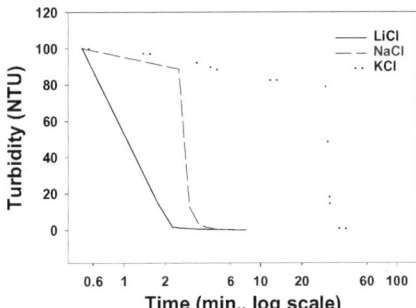

Figure 4: Turbidity data for suspensions containing 1.0 M salt

significant decrease in turbidity (settling of suspensions) by adding 0.01 M and 0.1 M salt. However, addition of 1.0 M salt resulted in rapid coagulation of the slurry, as can be seen from Fig. 4, with LiCl and NaCl being more effective coagulants than KCl. In addition, particle size measurements as a function of salt concentration showed that no coagulation occurred till a salt concentration of 1.0 M. At 1.0 M salt, all three salts caused coagulation of the slurry. However, LiCl was the most effective coagulant at this concentration, followed by NaCl and KCl. This is seen in Fig. 5, which shows a representative particle size distribution for each salt at 1.0 M. Colic et al. [7], also showed that LiCl and NaCl are more effective than KCl in coagulating silica slurries at high salt concentrations. From the above results, we can say that the increased removal rate due to salt addition is not due to any agglomerates that may be present in the slurry, as the lower salt concentrations do not cause any observable coagulation, and yet lead to significant enhancement in removal rate. That, along with the low roughness of all the polished samples, suggests that the enhanced wear rate is primarily due to increased contact between the surfaces participating in the polishing process. It

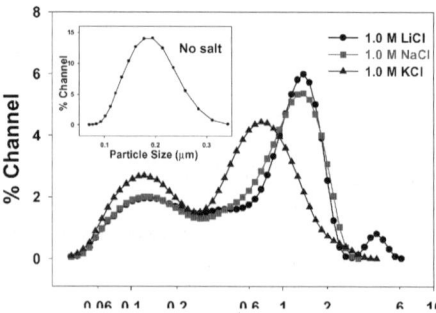

Figure 5: Particle size distributions for slurries containing 1.0 M salt. Inset: Original size distribution (without salt)

is well known that silica surfaces (wafer and particles) have a strong negative charge at alkaline pHs, and we have determined that the pad surface is also negatively charged under these conditions [8]. Due to addition of salt to the system, the alkali metal counterions compress the electrical double layer around the particles, wafer and pad surfaces, which reduces the electrostatic repulsion between them, thus increasing the effective contact between the particles, wafer and pad. Similar results have been reported by us earlier [8] in the measurement of frictional forces between a SiO_2 wafer and polishing pad, where we have shown that higher ionic strength solutions lead to higher friction and consequently higher removal of material from the wafer surface. This interpretation is also supported by the zeta potential, particle size and turbidity results, which show that at higher salt concentrations, KCl is less effective at reducing zeta potential as well as at coagulating the slurries. This explains the relatively smaller increase in removal rate in the slurries containing 1.0 M KCl, as compared to LiCl and NaCl. Chemical interactions between silica and the salts probably plays a role too [9], but its effect is overshadowed by the increase in the mechanical interactions, and is beyond the scope of the present study.

CONCLUSIONS

The effect of adding salt on polish rate of SiO_2 slurries was studied. From polishing rate measurements, it was shown that adding salt causes significant enhancement in removal rates. Surface roughness measurements showed that this increase was not due to any agglomerates formed by the addition of salt. Zeta potential and turbidity measurements showed that small salt concentrations cause decrease in zeta potential but no significant coagulation. Correlating these results with polishing rate data and previous studies on friction force measurements, it has been shown that the increased wear rate is primarily due to increased contact between the wafer, particles and pad, caused by the reduced electrostatic repulsion caused by the addition of salt to the slurries. The effectiveness of a particular salt in suppressing the surface charge on the interacting surfaces at high salt concentrations correlates with its effectiveness in enhancing the oxide polish rate. From these results, we can say that soft agglomerates formed by increasing the ionic strength of CMP slurries are not necessarily detrimental to the CMP process.

ACKNOWLEDGMENTS

The authors would like to acknowledge the financial support of the Engineering Research Center (ERC) for Particle Science and Technology at the University of Florida, the National Science Foundation NSF grant #EEC-94-02989, and the Industrial Partners of the ERC. Thanks are also due to J. Adler, G. Bassim and P. Singh for valuable discussions.

REFERENCES

1. J.M. Steigerwald, S.P. Murarka and R.J. Gutmann, *Chemical Mechanical Planarization of Microelectronic Materials*, John Wiley & Sons, New York (1997)

2. L.M. Cook, J. Non-cryst. Solids **120**, 152 (1990)

3. R.J. Hunter, *Introduction to Modern Colloid Science*, Oxford University Press, New York (1993)

4. R.W. O'Brien, D.W. Cannon and W.N. Rowlands, J. Colloid Interface Sci. **173**, 406 (1995)

5. M. Bielmann, U. Mahajan, R.K. Singh, D.O. Shah and B.J. Palla, Electrochem. Solid State Lett. **2** (3), 148 (1999)

6. M. Bielmann, U. Mahajan and R.K. Singh, Electrochem. Solid State Lett. **2** (8), 401 (1999)

7. M. Colic, M.L. Fisher and G.V. Franks, Langmuir **14**, 6107 (1998)
8. U. Mahajan, M. Bielmann and R.K. Singh, Electrochem. Solid State Lett. **2** (2), 80 (1999)

9. R. Iler, *The Chemistry of Silica*, John Wiley and Sons, New York (1979)

Dielectric and Metal CMP

Mat. Res. Soc. Symp. Vol. 613 © 2000 Materials Research Society

Planarization of Cu and Ta Using Silica and Alumina Abrasives - A Comparison

Y. Li[ae], **S. Ramarajan**[bc], **M. Hariharaputhiran**[bc], **Y. S. Her**[d] **and S.V. Babu**[bc]
Departments of Mechanical[a] and Chemical[b] Engineering
[c]Center for Advanced Materials Processing
Clarkson University, Potsdam, NY 13699
[d]Ferro Corporation, Penn Yan, NY

ABSTRACT

Experimental results on chemical mechanical polishing of copper and tantalum in the presence of different chemicals, such as $Fe(NO_3)_3$, H_2O_2 and glycine, using both silica and alumina abrasive particles are presented. The polish rates with alumina slurries vary in accordance with the hardness of the material being polished, suggesting a dominant role for a wear mechanism. However, polish rates with silica slurries do not relate directly to the hardness of the film, indicating a more complex removal mechanism.

INTRODUCTION

Abrasives play an important role during chemical-mechanical polishing (CMP) process not only to enhance the polish rate but also in determining the extent of dishing, erosion and defects such as micro-scratches. Alumina and silica powders are widely used as abrasives for Cu and Ta CMP[1,2]. While alumina is harder than silica [3], and hence can produce higher polish rates, slurry stability [4] and cost concerns favor the use of silica abrasives.

The use of alumina and silica abrasives for CMP of Cu and Ta results in different mechanisms of polishing in terms of the effects of hardness of abrasive particles and the material being polished. In this paper, polish rates of Cu and Ta with both alumina and silica abrasives in the presence of different slurry chemicals are presented. The pH effect on the polish rate is also discussed.

EXPERIMENTAL DETAILS

Chemical-Mechanical Polishing

Polishing experiments were carried out using a bench-top Struers DAP-V polisher and 3mm thick Cu and Ta disks (99.99% pure) with a cross sectional area of 7.5 cm^2. The table speed was set at 90 rpm and the disk holder was held stationary. The applied downward pressure was approximately 6.3 psi (41.4 kN/m^2). Alumina and fumed silica powders were used as the abrasives. The alumina particle density was varied by controlling the calcination temperature during production (at Ferro Electronics), which alters the particle porosity and crystalline morphology. The details of the characterization of these alumina particles have been discussed elsewhere [5]. Different amorphous fumed silica particles, Cab-o-sil series (from Cabot Co.) and Aerosil series (from Degussa Co.) powders, were used. The particle concentration in the slurries was maintained at 3wt% by weight unless mentioned otherwise and the slurry feed rate was 60ml/min for all experiments. To maintain a good dispersion, the slurry in the slurry vessel was stirred continuously with a magnetic stirrer during polishing. A Suba 500 polishing pad was used and was hand conditioned prior to each experiment using a 220 grit sandpaper and a nylon brush. The polish rate was determined from the weight loss of the disk after being polished for 3 minutes and the reported data were obtained by averaging over four experiments.

Nanoindentation Tests

The hardness of each disk was measured using a Nanomechanical test instrument from Hysitron Inc. A diamond Berkovich tip was used. The loading and unloading rates were set at 100 μN/s. To minimize the effect of time-dependent plasticity, the load was held constant at the peak value for 5 seconds and then unloaded.

RESULTS AND DISCUSSION

Copper CMP

Figure 1 shows polish rates of Cu increase with increasing bulk density of alumina particles above a threshold value. Compared to polishing with alumina slurries, lower polish rates of copper are obtained with slurries containing silica particles Cab-o-sil L-90 and Aerosil 130 in the presence of different chemicals, as shown in Figure 2. Little difference is seen in the performance of the two silica abrasives. While Cu polish rate of approximate 220 nm/min is obtained with DI water containing 3wt% α-alumina particles, Cu is not polished in DI water with either of the silica particles due to the hardness of hydrated silicas (400-500 kg/mm^2) being lower than that of alumina (~2000 kg/mm^2) [6].

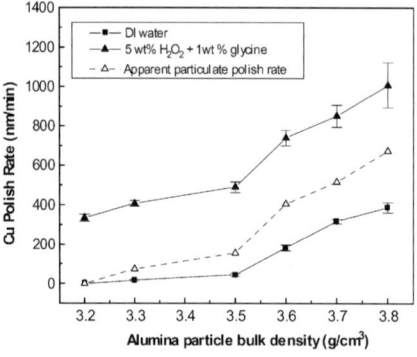

Figure 1. *Cu polish rates with alumina Particles.*

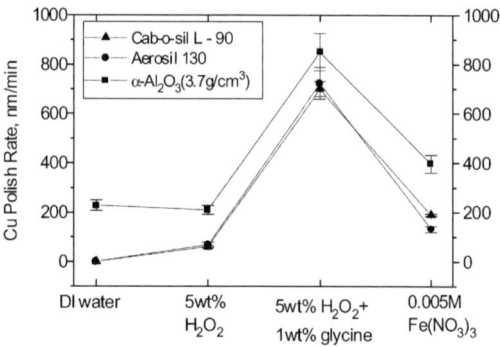

Figure 2. *Cu polish rates in different slurries.* (The lines are drawn only to show the trends.)

In the presence of 5wt% H_2O_2, the polish rate increases from that in DI water when silica abrasives are used but decreases slightly in the presence of alumina abrasives. The formation of a harder oxide film [7] in the presence of 5wt% H_2O_2 leads to a lower removal rate of Cu in alumina slurries, indicating a wear mechanism dominated by the mechanical properties of the abrasive particles and the film being polished. In silica slurries, however, the fact that the harder oxide, but not softer copper, is polished suggests a different approach involving the structure of the oxide film, the strength of the copper/oxide interface and possible chemical interactions with a reactive silica surface [7,8].

Adding 1wt% glycine, which forms a soluble complex with Cu^{2+}, increases copper polish rates dramatically in H_2O_2-based slurries, with both silica and alumina particles. The mechanism of Cu polishing in H_2O_2-glycine slurries has been addressed in a previous presentation [9] and will not be discussed here.

Cu polish rates in 0.005 M $Fe(NO_3)_3$ with and without abrasive particles are listed in Table I. A removal rate of approximate195nm/min is obtained in 0.005M $Fe(NO_3)_3$ without abrasives due to chemical dissolution and the relative motion between the Cu disc and the polishing pad. In the presence of alumina particles, the polish rate is a superposition of chemical dissolution and mechanical abrasion. However, using silica particles reduces Cu polish rate. Also, lower removal rates are obtained with Cab-o-sil LM-150 silica powders which have larger specific surface area than the Cab-o-sil L-90 silica powders. This is presumably caused by the adsorption of Fe^{3+} ions by the silica particles.

Table I. Cu polish rates in 0.005M $Fe(NO_3)_3$

Particle concentration, wt%		DI water	0.005 M $Fe(NO_3)_3$
Without abrasives		0	195 ± 22
α-Alumina (3.6 g/cm^3),	3wt%	150 ± 10	350 ± 15
α-Alumina (3.7g/cm^3),	3wt%	229 ± 25	400 ± 25
Silica Cab-o-sil L-90,			
	3wt%	0	189 ± 5
	6wt%	0	154 ± 10
Silica Cab-o-sil LM-150,			
	3wt%	0	164 ± 1
	6wt%	0	146 ± 2

Ultraviolet (UV) spectroscopy was used to track the Fe^{3+} ion concentration in the slurries with different silica particle concentrations. Silica particles were equilibrated in a known concentration solution of $Fe(NO_3)_3$ and filtered. The Fe^{3+} concentration of the clear solution was measured. The results, listed in Table II, show that, in the presence of 3wt% and 6wt% silica (Cab-o-sil LM-150) in the slurry, Fe^{3+} concentration in the filtrate is reduced by 41% and 57%, respectively. Iron ions are presumably adsorbed by the silica particles, reducing their effective concentration and, hence, the dissolution and polish rates. Similar results are obtained with other silica abrasives.

Table II. UV data of Fe^{3+} concentrations in slurries

Slurry	1mM $Fe(NO_3)_3$ solution without silica	1mM $Fe(NO_3)_3$ solution with silica LM-150	
		3wt%	6wt%
Fe^{3+} concentration, mM	1	0.59	0.43

Tantalum CMP

Figure 3 shows polish rates of Ta with 3wt% silica (Cab-o-sil L-90, Aerosil 130) and alumina particles in different as-dispersed slurries. Surprisingly, while the softer Cu is not polished in DI water with silica slurries, Ta is polished at a significant rate (~75 nm/min with Aerosil 130) in DI water. Alumina slurry produces a Ta polish rate of ~70 nm/min. There is also significant difference between the two silica abrasives, with the Cab-o-sil L-90 producing only a 37 nm/min polish rate, suggesting a profound impact of surface chemical interactions between the silica particles and Ta on the removal rate.

Furthermore, Ta polish rates decrease with the addition of H_2O_2, glycine as well as $Fe(NO_3)_3$. This is presumably caused by the formation of a harder oxide film on the Ta surface which is confirmed by the nanoindentation test, as shown in Figure 4.

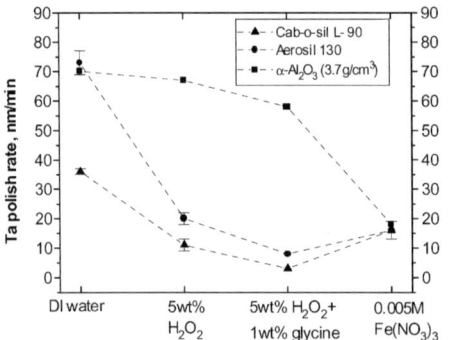

Figure 3. Ta polish rates in different slurries.   **Figure 4.** Miacrohardness of Ta disk.

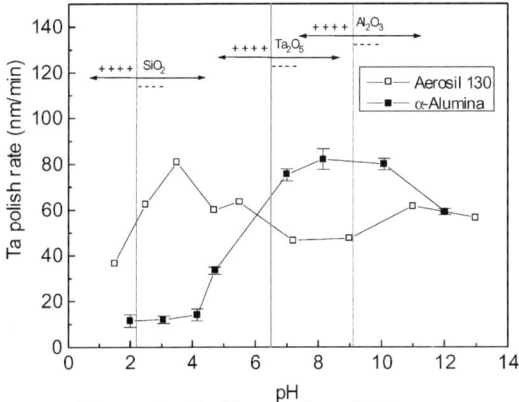

Figure 5. pH effect on Ta polishing rate.

Additionally, Ta polishing is strongly affected by the pH of the slurry, caused by the electrostatic interactions between the abrasive particles and Ta surface. The magnitude of repulsion or attraction between the abrasive particles and the Ta surface depends on the actual value of zeta-potential and the width of the electrical double layer around the surface. The slurry pH is adjusted using KOH and HCl. Figure 5 shows the polish rates of Ta as a function of slurry pH in DI water containing silica and alumina particles. The highest removal rates are obtained at pH values where the particles and the Ta surface are oppositely charged and experiencing electrostatic attraction. The effects of pH and ionic strength have been discussed in details in another paper [10]. Furthermore, Ta polishing in slurries in the presence of compound A shows great impact of pH on the removal rate, as listed in Table III. The polish rate increases dramatically at pH=12 with alumina abrasives.

Table III. Ta polish rates (nm/min) in compound A

	pH = 2.1	pH = 12.0
α-Al$_2$O$_3$	23 ± 1	256 ± 16
Cab-o-sil L-90	27 ± 4	30 ± 3

CONCLUSIONS

The polishing of Cu and Ta using sub-micron sized alumina abrasives is dominated by a wear process in which the removal of materials depends on the hardness of the abrasive particles and the film being polished. The formation of harder oxide films reduces the polish rates of both Cu and Ta in alumina slurries. However, the polishing with silica abrasives is complicated by the possible chemical interaction between the silica particles and the film being polished due to the active silica surface, suggesting a different mechanism. In any case, the pH strongly affects the polish rates.

ACKNOWLEGEMENTS

The authors would like to acknowledge project support from Intel Corporation, NY State Energy Research Development Authority, Ferro Corporation and the National Science Foundation (Grant CTS 9871264). The authors would also like to thank Rodel Inc., Cabot Corporation and Degussa Corporation for supplying polishing pads, abrasive powders and slurries.

REFERENCES

1. J.M. Steigerwald, S.P. Murarka, R. J. Gutmann, *Chemical Mechanical Planarization of Microelectronic Materials*, John Wiley & Sons, Inc. (1997).
2. M. Hariharaputhiran, Y. Li, S. Ramarajan and S. V. Babu, Chemical-Mechanical Polishing of Ta, *Electrochemical and Solid-State Letters,* **3** (2), (2000) pp.95-98.
3. A. A. IVAN'KO, *Handbook of Hardness Data*, Israel Program for Sicience Translations, (1971).
4. Luo, Q., Campbell, D. R., and Babu, S. V., *Stabilization of Alumina Slurry for Chemical-Mechanical Polishing of Copper, Langmuir,* 12(15), (1996) pp. 3563.
5. S. Ramarajan, M. Hariharaputhiran, Y.-S. Her and S. V. Babu, Hardness of Sub-Micrometer Abrasive Particles and Polish Rate Measurements, *Surf. Engg.,* **15** (4), (1999) pp.324-328.
6. M. L. Cook, Chemical Processes in Glass Polishing, *J. Non-Crystalline Solids,* **120**, (1990) pp.152-171.
7. Y. Li, M. Hariharaputhiran and S. V. Babu, Chemical-Mechanical Polishing of Copper and Tantalum with Silica Abrasives, submitted to *J. Mater. Res.*, May 2000.
8. R. K. Iler, *The Colloid Chemistry of Silica and Silicates*, Cornell University, Ithaca, New York, (1955).
9. M. Hariharaputhiran, J. Zhang, Y. Li and S.V. Babu, MRS Proceedings, vol. **566**, (1999) pp.129
10. S. Ramarajan, Y. Li, M. Hariharaputhiran, Y.-S Her and S. V. Babu, Effect of pH and Ionic Strength on The Chemical-Mechanical Polishing of Tantalum, *Electrochemical and Solid-State Letters,* **3**(5), (2000) pp.232-235.

Mat. Res. Soc. Symp. Vol. 613 © 2000 Materials Research Society

Chemical Wear of Cu CMP

Hong Liang, University of Alaska Fairbanks;
Jean-Michel Martin and Beatrice Vacher, Ecole Centrale de Lyon; and
Vlasta Brusic, Cabot Corporation

In this work, we used surface analysis techniques, such as a field-emission high-resolution analytical TEM, X-ray spectroscopy, and XPS to analyze abrasive particles after polishing. Results showed evidence of copper oxide (Cu_2O) in the polished slurry. However, there was no metallic crystalline copper detected. After comparing these data with the results obtained from our electro-chemical experiments, we propose two possible chemical wear mechanisms in Cu CMP.

Introduction

Chemical mechanical polishing (CMP) is a synergetic process that undergoes kinetic combination of three different components. Investigation in understanding CMP mechanisms has been focused on slurries, quality and defects on wafers, and pad behavior. Previous studies in effects of abrasive particles on CMP were done by extrapolating and estimating existing data. It was concluded that particles abraded a soft surface layer from wafer surfaces[1]. The metal (tungsten) CMP mechanisms have been accepted by many researchers as the passivation of a metal surface[2] but other researchers have found contradictory results[3]. These indicate the need for further investigation. The previous studies have been mainly focused on the chemical nature of CMP processes. Therefore, a broader view or aspects should be taken.

The interactions between abrasive particles and wafer surfaces during polishing result in changes on the contacted surfaces and in the chemical bonding types. Pioneer work has been done on tribochemical wear [4, 5, 6]. An understanding was established that chemical reactivity affects wear significantly. Under stress, chemical reactions proceed differently from those without stress[7]. Due to synergetic processes, CMP may posses more than one mechanism. In order to understand the effect of mechanical components on CMP, we characterized the reaction products by using the state-of-the-art surface analysis techniques. These techniques were proven being effective providing key information for understanding CMP.

Experimental

Copper CMP was performed on a modified bench-type disc-on-pad polisher. The frictional force, displacement, and applied pressure can be measured during polishing. A 2" square copper block was mounted on a self-rotating disk. The polished pad, used as IC1000, was mounted on a 4" rotating disc. Particles with a high pH (~ 10) in the deionized water were used to polish the copper block. The average pressure applied was 5 psi. The disk speed was 20 rpm. During polishing, slurry was collected for analysis.

A field-emission high-resolution analytical TEM and X-ray spectroscopy were used to analyze the abrasive particles after polishing. For comparison, abrasive particles before and after polishing were compared side-by-side. The X-ray photoelectron spectroscopy (XPS) was used for chemical analysis. The TEM and XPS have been used by Varlot et. al. at the *Ecole Centrale de Lyon Laboratory* for tribochemical studies of boundary lubrication[8]. The TEM (JEOL 2010 field emission microscope) was operated at 200kV accelerating voltage. The size of the electron beam was set at 2.4 nm. A holey carbon film was mounted on an aluminum grid. The carbon film was approximately 5 nm thick, which is particularly suitable for the high resolution TEM analysis.

Results

The SiO_2 particles before polishing were analyzed by using HR TEM and X-ray. The colloidal particles were deposited on a holey carbon film without being washed. They were then analyzed by using an X-ray spectroscopy. Figure 1 shows the results of the EDX spectra. In this figure, the filled area indicates the area covered by particles. The white lines indicate the area of outside the particle. The contimentation of aluminum is from the aluminum grit of the carbon film. The concentration of copper is an artifact from apertures in the microscope. This establishes a baseline for the analysis. In Figure 1, the existence of potassium K is visible in between 3 and 4 keV.

The colloidal particles were examined under TEM. The image and chemical mapping analysis results are shown in Figure 2. Figure 2a, positioned at the upper left corner of Figure 2, shows the aggregated colloidal particles. The open arrow corresponds to the light area in Figure 1, which is outside of the particle. The dark arrow corresponds to the filled arrow in Figure 1, which is inside of or on the colloidal particle. The silicon mapping is shown in Figure 2b, located at the upper right corner, where the higher darkness represents a higher concentration of silicon. Inside or on top of the particle, the concentration of silicon is apparently high. The potassium and copper were mapped and the results are shown in Figure 2c at the lower left corner and Figure 2d at the right. It is interesting that the two elements are homogeneously contributed inside and out of the silica particle. Potassium was a common addition as KOH to silica slurry. It is not surprising that there is no concentrated K on silica particles. Cu is from the artifact of the aluminum grit. Therefore, we expected that no concentrated copper would be found.

Polished silica particles were analyzed as shown in Figures 3 and 4. Figure 3 shows the results of the EDX spectra of slurry after polishing. Compare with Figure 1, there is a significant amount of copper present on and off silica particles. The arrows showing in Figure 3b, at the right of Figure 3, correspond to the white and filled areas in Figure 3a. Under TEM, the image and chemical mapping analysis were conducted and the results are shown in Figure 4. Figure 4a, positioned at the upper left corner of Figure 4, shows the silica mapping. The high concentration of silicon on the particles forms clearly the shape of aggregated colloidal particles. The oxygen mapping, as shown in Figure 4b, located at the upper right corner of Figure 4, shows high concentration inside the particles. This is due to the formation of SiO_2. The copper mapping shown in Figure 4c indicates the existence of copper inside and out of silica particles. Comparing with Figure 2d, the shape of silica particles is slightly distinguishable in Figure 4c after polishing. This indicates possible interactions between copper and silica particles. The potassium mapping is shown in Figure 4d (lower right corner of Figure 4.) The presence of potassium is similar to copper before and after polishing. It indicates the possible interaction between potassium and silica particles during polishing. Further chemical analysis of Cu and oxygen inside the slurry was conducted by using XPS. Results showed that there was no copper crystalline in the slurry. Most likely the copper was in the form of Cu_2O.

Discussions:

Previous results showed that during an electro-chemical test in SS-25, Cu dissolution rate with abrasion is 70-100 Å/min[9]. Cu corrodes at about 2 Å/min without any mechanical abrasion. According to Brusic[10], in the alkaline pH region of oxide polishing slurries, Cu passivates immediately when a pure copper is exposed to oxygen. The evidence of passivation is noticeable even with abrasion[9, 11].

According to electrochemical measurement, addition of mechanical abrasion increases the copper dissolution rate from 2 to larger than 70 Å/min. The effect of abrasion on the corrosion rate is significant. Our surface analysis results showed that there was no metallic particle nor large copper oxide worn debris detectable in the polished slurry. Instead, there was a dissolved form of copper, i.e., copper oxide (possibly Cu_2O) existing in the slurry. Therefore, the direct mechanical wear of copper is not likely. The mechanical energy applied onto the copper surface must have been dispersed into a surface layer or pure copper surface and the atomic copper was affected. There are apparently two possible polishing mechanisms.

The first possibility is to assume the passivation of the copper surface at alkaline pH region. The passivated film was formed quickly. Under the polishing action, the surface energy of oxide film was excited by mechanical interactions between the pad and abrasive particles. The bonding of copper oxide breaks down and therefore it dissolves into the slurry solution.

The second possibility is that the pure copper surface is generated during polishing. The surface energy is excited by mechanical interaction and the copper bonding breaks down. Copper atoms immediately oxidize in the slurry and Cu_2O remains.

Both mechanisms are chemically based (chemical wear) processes. This hypothesis will be further tested for future studies.

New colloidal SiO$_2$

Colloidal particles are deposited into a very thin carbon film supported itself by a holey carbon film, with aluminum grid

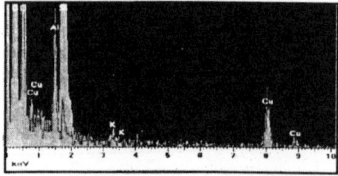

Figure n°1: EDX spectra corresponding areas (scaled on X-ray Al)

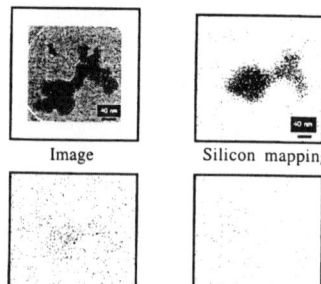

Image   Silicon mapping

Potassium mapping   Copper mapping

Figure n°2: Chemical cards and TEM Image

Polished colloidal SiO$_2$

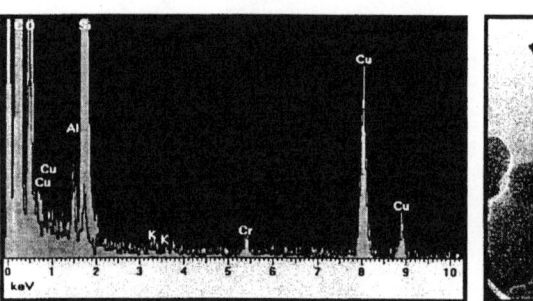

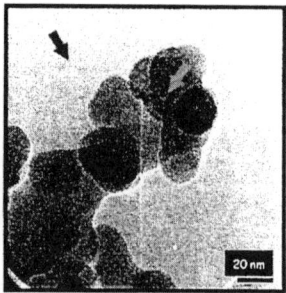

Figure n°3: EDX spectra corresponding areas (scaled on X-ray Al) and TEM image

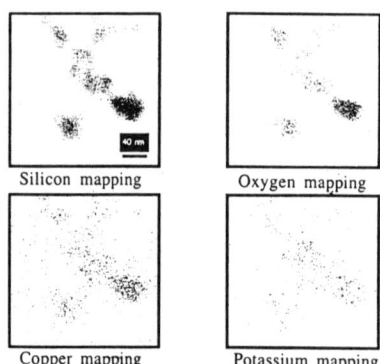

Silicon mapping   Oxygen mapping

Copper mapping   Potassium mapping

Figure n°4: Chemical cards

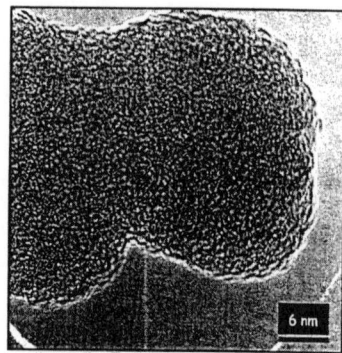

Figure n°5: HREM Image polished colloidal SiO$_2$

Conclusions:

Based on our direct observation of polished slurry particles, we conclude that the possible CMP mechanism is a chemically-based wear process. The atomic removal of surface material was made possible by mechanical stimulation, which took the form of either direct removal of the copper oxide film or of the copper metal surface.

Acknowledgments:

This work was partially supported by the President Special Fund Grant and Alaska Space Grant Program at UAF. The authors would also like to thank Ms. Sandra Boatwright for editing the text and Dr. Frank Kaufman from Cabot Corporation in providing polishing slurries.

References:

[1] J. Larsen-Basse and H. Liang, "Probable Roles of Abrasive Particles in W CMP," Wear, 233-235, 647-654, 1999.

[2] F. B. Kaufman, D. B. Thompson, R. E. Broadie, M. A. Jaso, W. L. Gutherie, D. J. Pearson and M. B. Small, "Chemical-Mechanical Polishing for Fabricating Patterned W Metal Features as Chip Interconnects," J. Electrochem. Soc., 138, 3460, 1991.

[3] D. Stein, d. Hetherington, T. Guilinger, and J. Cecchi, "In Situ Electrochemical Investigation of Tungsten Electrochemical Behavior during Chemical Mechanical Polishing," J. Eletrochem. Soc., 145, 3190, 1998.

[4] J. M. Martin, Th. Le Mogne, M. Boehm, and C. Grossiord, "Tribochemisty in the Analytical UHV Tribometer," Trib. Int., in press.

[5] J. M. Martin, Th. Le Mogne, C. Grossiord, and Th. Palermo, "Tribochemistry in the Analytical UHV Tribometer," Trib. Let., 3, 87-94, 1997.

[6] J.M. Martin, "Antiwear Mechanisms of Zinc Dithiophosphate: a Chemical Hardness Approach," Tribology Letters, Vo. 6, 1-8, 1999.

[7] J.T. Dickinson, N.-S. Park, M-W. Kim, and S.C. Langford, "A Scanning Force Microscope Study of a Tribochemical System: Stress-Enhanced Dissolution," Trib. Lett. 3, 69-80, 1997.

[8] K. Varlot, J. M. Martin, C. Grossiord, B. Vacher, and K. Inoue, "A Dual Analysis Approach in Tribochemistry: Application to ZDDP/calcium Borate Additive Interactions," Trib. Letters.

[9] B. Brusic, D. Scherber, F. Kaufman, R. Kistler, and C. Streinz, "Electrochemical Approach to Au and Cu CMP Process Development," The Electrochemical Society Proceedings, V. 96-22, 176-185, 1997.

[10] V. Brusic and H. Liang, Private Conversation.

[11] V. Brusic, M. A. Frisch, B. N. Eldridge, F. P. Novak, F. B. Kaufman, B. M. Rush, and G. S. Frankel, "Copper Corrosion with and without Inhibitors," J. Electrochem. Soc., Vol. 138, No. 8, 2253-2259, 1991.

Poster Session:
Dielectric and Metal CMP

Mat. Res. Soc. Symp. Vol. 613 © 2000 Materials Research Society

A NOVEL SINGLE STEP LAPPING AND CHEMO-MECHANICAL POLISHING SCHEME FOR ANTIMONIDE BASED SEMICONDUCTORS USING 1 μm AGGLOMERATE-FREE ALUMINA SLURRY

P.S. DUTTA*, R.J. GUTMANN*, D. KELLER**, L. SWEET***
*Department of Electrical, Computer and Systems Engineering,
 Rensselaer Polytechnic Institute, Troy, NY, USA
**Universal Photonics Inc., Hicksville, NY, USA
***Baikowski International Corp., Charlotte, NC, USA

ABSTRACT

A novel approach for a single step lapping and chemical-mechanical polishing of antimonide-based III-V compounds using agglomerate-free alumina slurries is presented. Relatively high removal rates, minimal scratching, and low surface roughness have been obtained. The effects of slurry preparation cycle on the slurry properties and chemo-mechanical polishing results are discussed.

INTRODUCTION

Traditionally, wafer polishing of elemental and compound semiconductors is performed in multiple steps. In the first step, the sawed wafer is made parallel on both sides using a large diameter abrasive (15-20 μm) on a hard pad. Subsequently, the highly parallel wafer is polished with decreasingly lower abrasive size particles (down to 0.01 μm) on medium hardness pads and soft pads. This strategy is well established for most of the elemental and binary compounds such as Si, Ge, GaAs, InP etc [1,2]. For relatively soft or fragile compounds such as III-V antimonides, II-VI semiconductors and recently developed ternary and quaternary III-V substrates [3-5], usage of large diameter lapping abrasives introduces permanent cracks in the wafers. Especially for the ternary and quaternary alloys, large lattice stress exist [6] and the wafers easily crack during the lapping process. Availability of high quality ternary and quaternary substrates in large scale will open up new avenues for devices employing novel band gap engineering strategies. Until recently, bulk ternary and quaternary substrates could not be grown without cracks. Growth of crack- free ternaries GaInSb, GaInAs and quaternary GaInAsSb crystals have now been made possible using novel approaches [3-5,7] and the feasibility for large scale production has been demonstrated. However, commercial development of ternary and quaternary substrate technology not only depends on the improvements in the crystal growth techniques, but also on the substrate slicing and polishing procedures. Substrate preparation issues have been addressed in this paper.

Initial work on lapping and polishing GaInSb poly- and single crystalline materials using conventional procedures (for elemental and binary compounds) led to excessive breakage of wafers. Multi-component alloys are in general fragile and care should be taken during the wafer preparation process. Cracking of wafer usually takes place when high mechanical force is applied during polishing and/or by using large diameter abrasive particles (> 5 μm in this case). The mechanical cracks are generated at the surface and then rapidly propagate to the entire bulk of the wafer. In this work, it has been demonstrated that the cracking can be avoided by using smaller abrasive particles (~ 1 μm) for lapping. However, the lapping and polishing times are increased enormously unless agglomerate- free slurries are used. Prolonged lapping and polishing cycles can also lead to mechanical damage of the wafers.

This paper presents a novel approach for preparing damage free antimonide-based III-V substrates of GaSb and GaInSb using the Baikalox CR- series of agglomerate- free alumina slurries. Rapid removal of material in a damage- free fashion has been evidenced with slurries containing abrasive particles of 1 μm and less. The surface finish obtained using these slurries is close to that of device-grade silicon. The high removal rate exhibited by these slurries is attributed to the agglomerate- free nature resulting from special preparation procedures.

EXPERIMENTAL DETAILS

Single- and poly- crystalline wafers of GaSb and $Ga_{1-x}In_xSb$ ($0.05 < x < 0.25$) were sliced from in-house grown bulk crystals [3-5] using a Princeton Scientific wire saw (model WS-22) with boron carbide (14 μm abrasives)-glycerine slurry. The sliced wafers exhibited the usual saw ridges, but no mechanical cracks. GaSb was used as a base-line comparison for the ternary substrates. For lapping and polishing, the samples were attached to a holder using crystal bond (wax) from MR Semicon Inc. A variety of alumina-based slurries have been evaluated along with several different types of soft and hard pads (such as Chemo-Tex, Nylon, Velvet from South Bay Technology and Rodel IC-1400). Buehler Micro-polish and Malvern Multipol polishing systems were used. In the Buehler Micropolish unit, the slurry is contained in a shallow enclosure with a fixed pad. The sample holder is attached to an arm which performs an oscillatory rotation. In the Multipol unit, the sample rotates along with the pad (in the opposite direction) while the slurry is continuously dispensed. Typical sample rotation speed employed in this work varied in the range of 20 – 50 rpm. Between subsequent polishing steps, the sample along with the holder was cleaned in de-ionized water, kept inside an ultrasonic bath and dried with nitrogen gun. After polishing, the wafers were inspected under optical microscopy. Atomic Force Microscopy (AFM) measurements have been used to evaluate the micro-roughness of the polished surfaces. The AFM scans were performed using a Nanospec AFM system in the tapping mode. In most of the cases, the scratches mentioned here are not seen with optical microscopy.

EFFECT OF FORCED CONVECTION ON SLURRY PROPERTIES

The hydrodynamic fluid motion during slurry preparation has been found to influence slurry viscosity and agglomeration. For the same weight content of abrasives, slurries prepared using different mixing schemes possess significantly different characteristics. In the present study, water-based alumina slurries with particle sizes ranging from 0.05 – 1 μm has been prepared by mixing powder abrasives with water. High shear mixing during slurry preparation has been found to be critical for avoiding agglomeration [8]. Slurries prepared using this scheme exhibited higher viscosity (for the same weight percentage of the abrasive particles) and no decantation (slurry settling) after several months. On the other hand, the same slurries prepared using simple mixing showed agglomeration and abrasive separation within a few hours of preparation and possess low viscosity. The weight percentage of abrasive in the slurries was in the range of 10 – 20%. The viscosity and suspension properties of the water-based alumina slurries could be altered by addition of glycerine.

Several batches of agglomerate-free alumina slurries (the Baikolox CR- series) with abrasive sizes of 1, 0.3 and 0.05 μm were investigated. The abrasive weight percentages in these slurries were in the range of 10 – 14%. These high viscosity slurries contain some suspending agent along with glycerine, water and abrasive. The slurries remain suspended for a relative long period of time (few weeks to several months). They were used without any further modification in our experiments. The agglomerate-free nature of these slurries and superior suspension properties result from the intricate slurry preparation procedures including high shear mixing [9].

POLISHING EXPERIMENTS AND RESULTS

The slurry properties have been found to influence the removal rate and micro-scratching. The removal rates for the agglomerate-free slurry was found to be at least four times higher than the same slurry containing agglomerates. The high removal rates eliminates the need for lapping using large abrasive particles ($12 - 30$ μm), which gives rise to deep scratching and sub-surface damages and hence is of utmost importance for fragile alloys such as GaInSb and GaInAsSb. Typical removal rates with the Baikalox 1 μm slurry was found to be similar to that obtained using an agglomerated alumina slurry with $12 - 14$ μm abrasives. Apart from the high removal rates, the agglomerate- free 1 μm slurry gives rise to a mirror smooth high quality surface as opposed to the usual lapping slurries. Hence, the number of polishing steps to obtain a device-grade surface reduces considerably. In our experiments, we have been able to reduce the total number of lapping and polishing steps to merely two: the first step was performed using the Baikalox 1 μm slurry followed by the final step with either a colloidal alumina or silica slurry with $0.02 - 0.05$ μm abrasives. The entire process of lapping and polishing is thus simplified and yields superior result compared to the conventional multi-step process.

With the agglomerate- free slurries, the removal rates are less dependent on the pad material. Surprisingly, a softer pad such as velvet polished at a faster rate than some of the hard pad such as Rodel IC-1400. The influence of pad material and structure has been found to be critical in maintaining flatness. The hard pads such as Chemo-Tex and IC-1400 are suitable for superior flatness; however, they lead to a high density of micro-scratches as shown in the AFM image in Fig. 1. This has been solved by employing a final short polishing step on soft pads like velvet or nylon, which eliminates most of the microscopic scratches and is essential for obtaining defect free surfaces with atomic flatness. The polishing cycle on the soft pad needs to be short in order to maintain wafer flatness. Typical polishing times on velvet pad were in the range of $2 - 5$ minutes. The AFM image of GaSb single crystal polished using the IC-1400 pad followed by velvet (with the Baikalox 1-μm alumina slurry) is shown in Fig. 2. The RMS roughness is close to that obtained with an elemental semiconductor such as silicon.

The effect of slurry agglomeration on the surface characteristics of $Ga_{0.85}In_{0.15}Sb$ is shown in Figs. 3 and 4. These wafers were polished using agglomerated (Fig. 3) and agglomerate- free (Fig. 4) 1-μm alumina slurries. The pads used were Rodel IC-1400 followed by velvet (as in Fig. 2). Multiple deep scratches could be seen in the AFM image of Fig. 3. Moreover, the removal rate was found to be at least 2-3 times slower than what was observed with the agglomerate-free slurry (Baikalox).

In the present study, device-grade atomically flat surfaces of both GaSb and GaInSb have been obtained using an optimized three step lapping and polishing cycle. In the first step, the agglomerate- free 1-μm alumina slurry was used to remove slicing ridges on the wafer until a flat surface is obtained. This step is usually performed on a rigid pad such as Rodel IC-1400 or Nylon. Typical lapping times depends on the non-uniformity of the wafer thickness. In this study, the lapping time was in the range of $30 - 60$ minutes. The lapped wafer exhibit a mirror smooth surface, but possess multiple fine scratches. To eliminate the fine scratches, a short polishing cycle ($\sim$ 5 minutes) is performed using the same 1 μm slurry on a velvet pad. After this step, no visible scratches could be seen on the wafer surface. However, shallow scr-

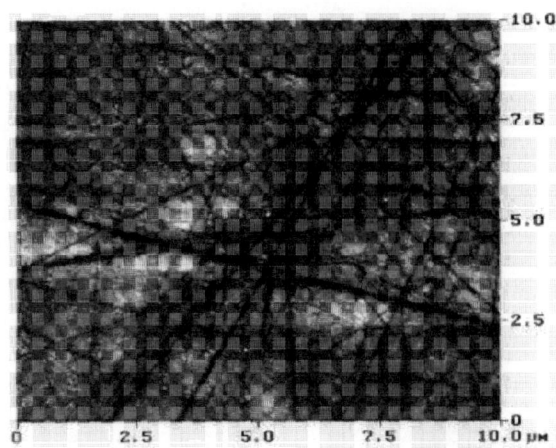

FIG. 1 AFM image of GaSb polished with Baikowski-Universal Photonics alumina (1 μm) slurry on Rodel IC-1400 pad. The RMS roughness is 2.62 nm.

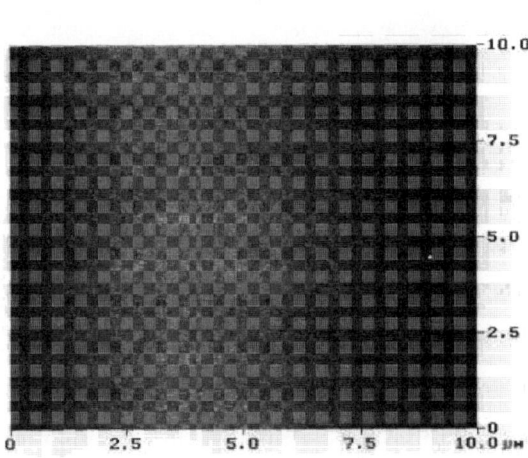

FIG. 2 AFM image of GaSb polished with Baikowski-Universal Photonics alumina (1 μm) slurry on Rodel IC-1400 followed by velvet pads. The RMS roughness is 0.25 nm

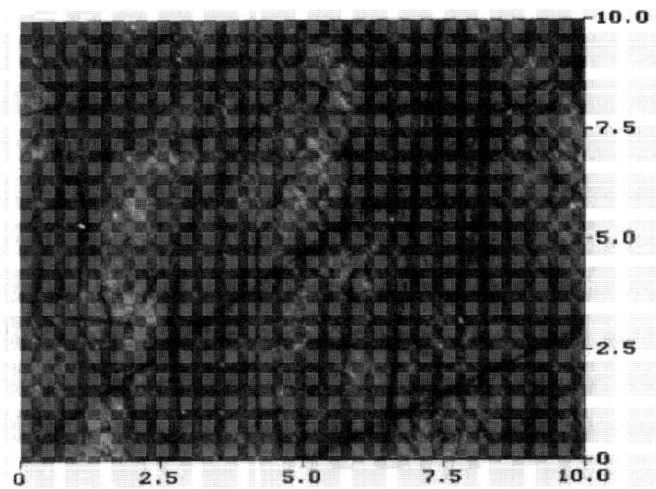

FIG. 3 AFM image of $Ga_{0.85}In_{0.15}Sb$ polished using an agglomerated alumina (1 μm) slurry on Rodel IC-1400 followed by velvet pads. The RMS roughness is 1.6 nm.

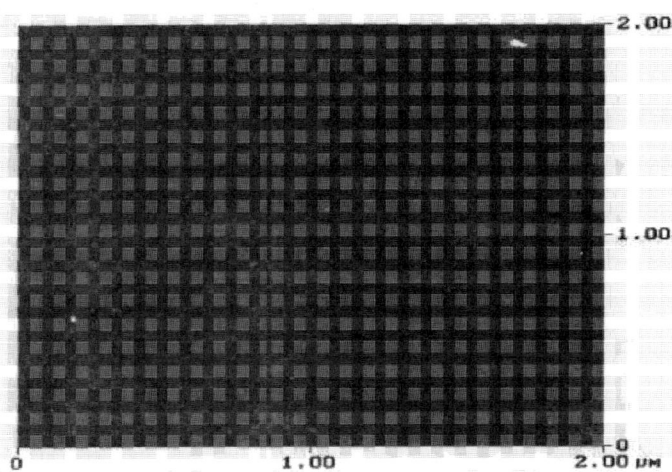

FIG. 4 AFM image of $Ga_{0.85}In_{0.15}Sb$ polished using Baikowski-Universal Photonics alumina (1 μm) slurry on Rodel IC-1400 followed by velvet pads. The RMS roughness is 0.1 nm.

-atches could be still seen in the AFM images (similar to that in Fig. 2). To eliminate the fine scratches, we explored a variety of water and oil based colloidal silica and alumina slurries with pH in the range of 3.5 to 10.5. The abrasive particle sizes in these slurries were in the range of 0.02 – 0.05 μm. The slurries with high pH such as Glanzox (colloidal silica with pH: 10.5) was found to be suitable for achieving an atomically smooth surface with no scratches. The final step is usually performed for 5 minutes on a velvet pad.

The lapping and polishing strategy developed in this work is quite general and applicable to other multi-component alloys. Similar results have been obtained in preliminary work on chemical mechanical planarization of liquid phase epitaxial grown AlGaAsSb/GaSb and GaInAsSb bulk substrates polishing experiments using agglomerate-free diamond-based and alumina-based slurries, respectively.

Post-CMP cleaning is critical in obtaining atomically flat surfaces. For blanket wafer surfaces, brush cleaning is widely used [2]. From AFM analysis of polished wafer surfaces, it has been observed that some of the fine abraded or slurry particles do not detach even after brush cleaning. A simple way of cleaning the polished surfaces using DOWFAX surfactant on a velvet pad has been found to be effective [8].

SUMMARY

A new technique for lapping and polishing fragile and soft multi-component semiconductor alloys has been developed using agglomerate- free slurries. Damage free substrate preparation of GaInSb has been possible using this methodology. Atomically flat surface of GaSb has been achieved by using agglomerate- free polishing slurries. It has been found that good slurry dispersion eliminates scratches and sub-surface damage, while increasing the removal rate drastically. High shear mixing has been found to be critical for obtaining agglomerate free slurries.

REFERENCES

1. J.E. Steigerwald, S.P. Murarka and R.J. Gutmann, "Chemical-Mechanical Planarization of Microelectronic Materials," J. Wiley (1997)
2. S.H. Li and R.O. Miller, "Chemical Mechanical Polishing in Silicon Processing", Semiconductors and Semimetals Series, Vol 63, Academic Press (2000)
3. P.S. Dutta and T.R. Miller, Patent Application Filed by Rensselaer Polytechnic Institute, Troy, NY (October 1999)
4. P.S. Dutta and T.R. Miller, J. Electronic Materials (in press)
5. P.S. Dutta, T.R. Miller, G.W. Charache and R.J. Gutmann, J. Crystal Growth (in press)
6. K.J. Bachmann, F.A. Thiel and H. Schreiber, Jr., Progress in Crystal Growth and Characterization 2, 171 (1979)
7. P.S. Dutta, A.G. Ostrogorsky and R.J. Gutmann, in: Proc. NREL Conf. On Thermophotovoltaic Generation of Electricity, Denver, CO, Ed. T.J. Coutts, J.P. Benner, C.S. Allman, AIP Conf. Proc. Vol. 460, New York, (1998) p. 227
8. P.S. Dutta and R.J. Gutmann, Proc. of the Fifth International Conference on CMP-MIC, Santa Clara, CA (2000) p. 441
9. S.H. Li, B. Tredinnick and M. Hoffmann, in Chapter 5 "Chemical Mechanical Polishing in Silicon Processing", Semiconductors and Semimetals Series, Vol 63, Academic Press (2000) p. 146, 148

Mat. Res. Soc. Symp. Vol. 613 © 2000 Materials Research Society

MULTI-LEVEL DAMASCENE PROCESS DEVELOPMENT: ALUMINUM CMP

David A. Hansen[a], Gerry Moloney[a] and Alex Reyes[a]
Cybeq Nano Technologies, San Jose, CA
[a]Present Affiliation: Multi Planar Technologies, Inc, San Jose, CA

ABSTRACT

The purpose of this work was to investigate integration of aluminum multi-level damascene devices with the CMP module. Traditionally aluminum is used in the Metal-1 level of device manufacturing and Reverse Ion Etching (RIE) is used to remove the aluminum over layer. However for <0.25-μm device rules RIE is not a method of choice due to incomplete removal of the over layer leaving stringers that cause shorting, as well as poor With-In Wafer Non-Uniformity (WIWNU). The result is loss of device yield and process problems for BEOL modules. Integrating aluminum device wafers into the CMP module has its own drawbacks such as immature consumables, i.e., slurry and polishing pads, as well as ease of scratching the soft aluminum interconnect structures. Hence the basis of this work is to highlight the main issues that impede the integration of the aluminum CMP process. Specifically we investigated the origin of deep aluminum scratches and the effect of not completely removing the residue barrier material, which can cause shorting and poor electrical performance. It is argued in this work that the observed deep scratches in the aluminum material are due to the polished debris emanating from the titanium glue layer. Recommendations are made to help to reduce this effect by design modification to the die layout on the patterned wafer and using electrical testing methods to help ascertain the minimum Over Polishing (OP) time required in order to ensure maximum die yield.

INTRODUCTION

For an ideal multilevel aluminum damascene process, aluminum CMP would remove the entire aluminum over-layer with no dishing of the aluminum bond pads and electrical wiring with no oxide loss. In other words the plug and line height would be defined by the initial etch depth with no detects. Physically however the barrier material has a removal rate that differs from the aluminum material. This causes dishing of the bond pads and electrical lines. The extent to which this occurs is proportional to the local device dimensions and polishing non-uniformity. Hence the key CMP metrics to control the dishing and oxide erosion are the WIWNU and the stop time or end pointing of the CMP process. However linked to both the dishing and oxide erosion are the requirement that no residue barrier material remain after the CMP process is complete. One quick method that is used often to determine when the barrier residual is completely polished off the wafer is by using an optical microscope. The problem with this method however is that for residual metal thickness of ~100-Å the material cannot be observed visually and a more sophisticated method of detection is required such as a SEM. The issue with this technique is that it is a destructive technique. A non-destructive method of probing for residue material after CMP is electrical testing of the aluminum interconnect structure in each die.

We have shown in recent work [1] that an increase in the scratch density on the aluminum is observed during runs in which only DI is applied during the polishing of the aluminum wafer. In addition it was argued that if no pad dressing was carried out that the aluminum particles that were left behind tended to agglomerate into larger particles, which if left on the pad during polishing would create large scratches.

In this paper we describe a different mode for the creation of deep scratches then we just described. In particular, we have carried out a study to determine the origin of deep scratches due to the CMP planarization process by using two different die layouts and to plot the OP time as a function of bridge electrical performance so as to determine the minimum OP time required to ensure that all residual material is removed.

EXPERIMENTAL

For dishing of large aluminum feature sizes, characterization was carried out on 50-μm bond-pad damascene wafers. Damascene wafers were polished by different amounts of time and the topography across the damascene wafers was characterized with a High Resolution Profiler (HRP). The polishing tool employed throughout this work was the Cybeq Nano Technologies IP-8000 CMP tool. A single step slurry and concentric grooved polyurethane pad with a closed cell foam layer were employed to polish all wafers in this work. The abrasive particle used is the slurry was alumina. All polish and percent over polish times reported in this work refer to that of the transition from 100% polishing of the aluminum to the barrier film, which was carefully monitored with the End Point Detection (EPD) system see figure 1.

Experimental Results

One damascene lot that we planarized did not have die extending to the edge of the wafer, which we call DAMA-1. When planarizing these wafers using a down force of 5-psi we observed deep scratches that ranged from 500-1500-Å deep just as we made a transition to the Ti glue layer. However if we used a damascene wafers with die extending to the edge of the wafer, which we call DAMA-2, we did not observe deep scratches. An example of this is shown in Figure 2-A and 2-B. In this figure 2A is the result form polishing DAMA-1 and 2B from polishing DAMA-2. As can be seen there is a clear difference between the two wafer types in regards to scratches. DAMA-1 has them and DAMA-2 does not. These results are consistent for the two different lots of wafers. The major difference between the two lot types was large non-structure areas of Ti on the wafer in which the die did not extend to the edge of the wafer. it was observe in the process development stage of this work that polishing blanket Ti wafers resulted in increasing scratch size with increasing down force. Using lower down forces to polish blanket Ti wafers resulted in smaller or no observed scratching of the Ti film. Hence for all work carried out here we used a low down force, which resulted in smaller or no observed scratching. In addition we are recommending that device designers extend die to the edge of the wafer so that creation of large residual Ti particles are reduced and hence reduced scratching of the aluminum structure.

A 4-μm line width with a 50-% density structure was planarized using different over-polish times for the characterization of oxide erosion and the corresponding aluminum dishing and recess. In figure 3 we plot several HRP scans as a function of the OP times. Notice that the y-axis in these plots map out the amount of the aluminum and oxide that is removed during the planarization step. Notice that the major loss of aluminum in the small line structure is due to the erosion of the oxide material with a secondary loss due to the actual dishing of the aluminum lines. In figure 4 we plot the material removed from the 4-μm line structure as a function of the over-polish time. As may be seen from the graph the oxide erosion increases linearly with increasing over-polish time whereas the aluminum dishing saturates after an over-polish time of about 20-%. Hence the main effect for the loss of aluminum material in small lines is through oxide erosion. We used an optical microscope to ensure that the residual barrier material was completely removed of all device structures. As mentioned above for film thickness of ~100-Å the titanium material is invisible optically. We will return to this topic in the next section and argue that by using electrical testing methods we may be able to determine the amount of OP required to remove all residual materials.

A 50-µm-bond pad structure was planarized using different over-polish times for the characterization of oxide erosion and the corresponding aluminum dishing and recess. In figure 5 we plot several HRP scans as a function of the OP times. Notice that the y-axis in these plots maps out the amount of the aluminum and oxide material that is removed during the planarization step. Notice that the major loss of aluminum in the bond pad structure is due to the dishing of the bond pad with a secondary loss due to erosion of the oxide material, which leads to a reduction of planarity, see figure 5 (D). In figure 6 we plot the material removed from the 50-µm-line structure as a function of the over-polish time. As may be seen from the graph the amount of aluminum loss increases linearly with increasing over-polish time. Hence the main effect for the loss of aluminum material in the large structure areas is through dishing and recessing of the bond pads. This is in contrast to small feature sizes in which aluminum loss is through oxide erosion. Typically for damascene wafers etch deep is between 0.5-0.7-µm depending on the design rules. Hence the corresponding aluminum material thickness if perfectly planarized is between 0.5-0.7-µm. Device integration rules generally allow a 10-% metal loss of the initial etch. Hence the optimized process for damascene device structures will have an allowable aluminum material loss of between 500-700-Angstroms depending on the etch depth. In this work this would correspond to an OP between 10-20-%. Note here that we have not measured quantitatively if all residual barrier material has been removed. As mentioned above we used optical methods to determine if all residual metal was removed. Hence we used electrical measurements to quantitatively determine if all residual material has been removed.

In figure 7 we plot the bridge resistance as a function of OP. As may be seen from figure for OP less than 30-% the measured die on the wafer are shorting. These results indicate that the minimum OP to ensure maximum yield is 30-%. Most semiconductors manufacturing specifications target dishing of bond pads of 500-Å to 700-Å as mentioned above. The reason for this specification target is so BEOL modules do not have integration issues. Hence, it is recommended that in order to meet these specifications that two-step polishing slurry be developed similar to that created for copper damascene CMP integration. This two-step slurry would have the first step remove the Al over layer and then use either high selective slurry or 1:1:1 slurry to remove the barrier. These schemes would reduce dishing and oxide loss and depending on the chemistry reduce scratching issues.

CONCLUSIONS

We have shown that we can obtain good CMP process results for Al damascene wafers. We have indicated that the observed deep scratches may be attributed to Ti debris from the CMP process. The deep scratches were observed on damascene wafers that do not have die extending to the edge of the wafer and not for damascene wafers that have die extending to the edge of the wafer. It has been suggested that device designers extend die to the edge of the wafers so that large non-structure areas of blanket Ti not exist. This would help reduce Ti particles that cause scratches. It was also shown that for OP between 10-20-% that the CMP performance was good. However on plotting the bridge resistance as a function of OP is was found that a minimum of OP 30-% is required in order not to lose die to shorts. It was recommended that two-step slurry be formulated similar to copper CMP integration. This two-step slurry would have the first step remove the Al over layer and then use either high selective slurry or 1:1 slurry to remove the barrier. This scheme would reduce dishing and oxide loss and depending on the chemistry reduce scratching issues.

REFERENCES

[1] J. Hernandez, P. Wrschka, Y. Hsu, T. -S. Kuan, G.S. Oehrlein, H.J. Sun, D.A. Hansen, J. King and M. Fury, *J. Electrochem. Soc.,* **146**, 4647 (1999).

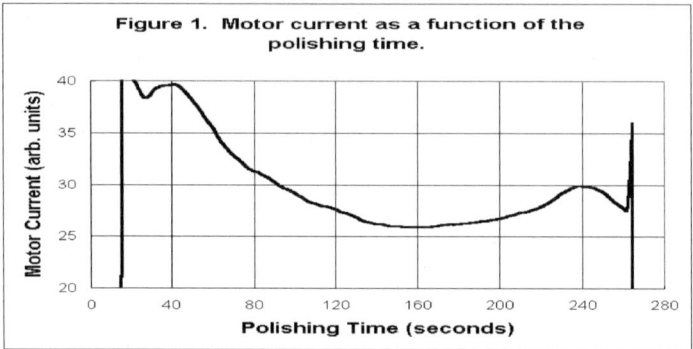

Figure 1. Motor current as a function of the polishing time for planarizing Al damascene wafer.

   (A) DAMA –1   (B) DAMA-2
Figure 2. (A) Corresponds to a wafer in which the die do not extend to the edge of the wafer and (B) to a wafer in which the die do extend to the edge of the wafer. It was observed in this work that wafer type (A) had scratching issues whereas (B) did not.

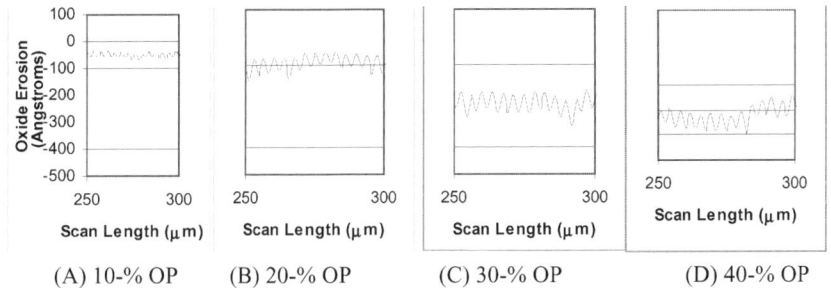

(A) 10-% OP   (B) 20-% OP   (C) 30-% OP   (D) 40-% OP

Figure 3. (A)-(D). Detailed HRP scans of 4-μm line structure as a function of percent over-polish. As may be seen increased polishing results in increased oxide erosion and dishing.

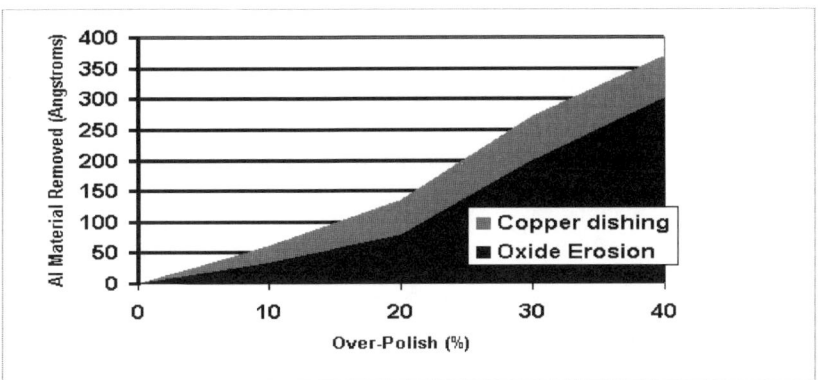

Figure 4. Al and oxide material removed as a function of OP for 4-μm line structure.

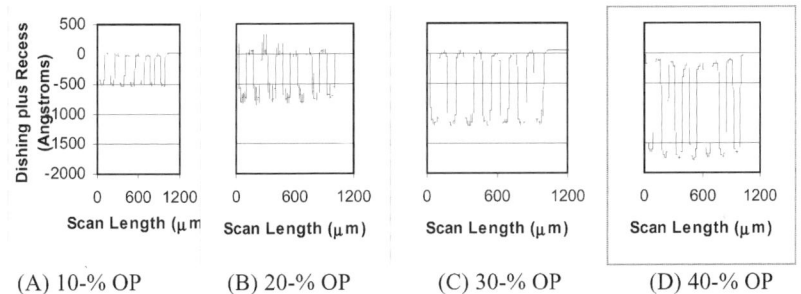

(A) 10-% OP   (B) 20-% OP   (C) 30-% OP   (D) 40-% OP

Figure 5 (A)-(D). Detailed HRP scans as a function of percent OP. As may be seen increased polishing results in increased dishing and recessing as well as loss of planarity.

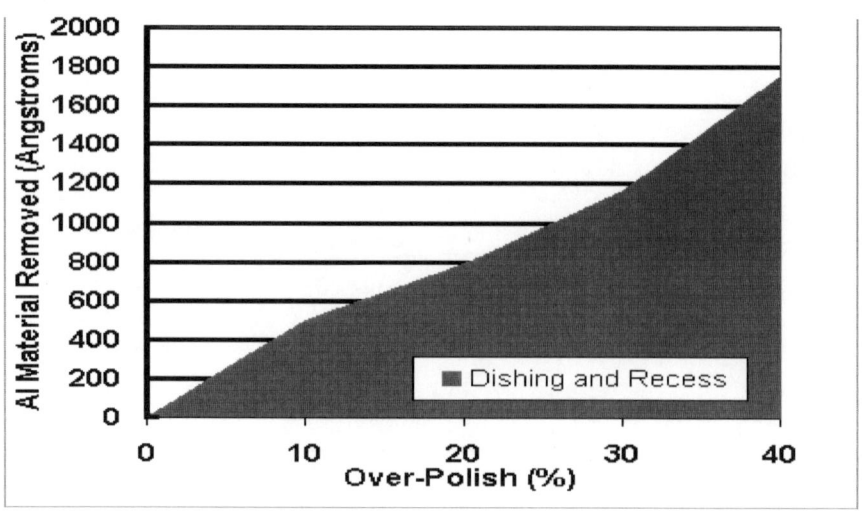

Figure 6. Al material removed as a function of OP for a 50-μm-bond pad.

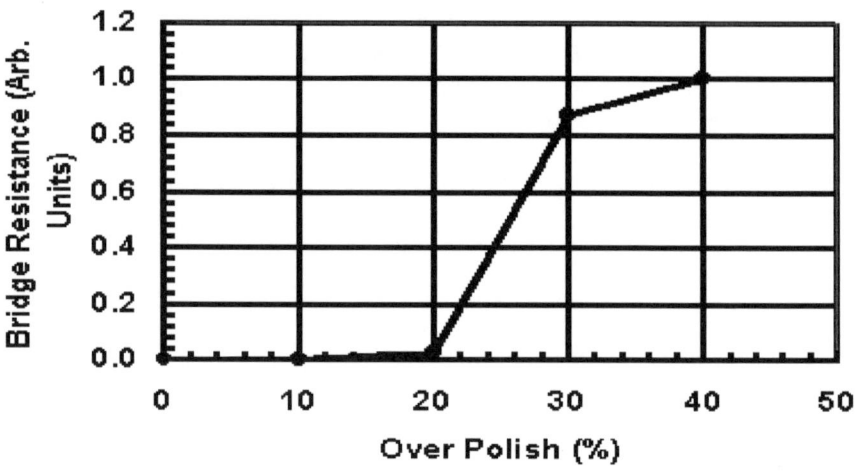

Figure 7. Bridge resistance as a function of OP for 0 0.4-μm bridge structure. Below an OP of 30-% the bridge resistance has shorts in the wiring on the level that had previously been CMP.

Poster Session:
Process Integration and
Manufacturability

Mat. Res. Soc. Symp. Vol. 613 © 2000 Materials Research Society

IMPROVEMENT OF WAFER EDGE PROFILE AND CMP PERFORMANCE THROUGH THE FLOATING HEAD DESIGN

Huey-Ming Wang*, Gerry Moloney*, and Mario Stella, Sesinando DeGuzman
Cybeq Nano Technologies, 45 E. Plumeria Drive, San Jose, CA 95134
*Present Address: Multi Planar Technologies Inc., San Jose, CA 95134

ABSTRACT

The dependence of IC fabrication on the Chemical Mechanical Planarization (CMP) process increases as the device features go down to 0.25 micron or beyond. Due to the tighter CMP process spec, it is very important to reduce the within wafer non-uniformity (WIWNU%) to achieve higher process yield. The symmetrical increment of linear velocity at wafer edge is not sufficient to change wafer edge profile by breaking the matched speed rule. A better solution is through the change of head design for a fixed platen from the polisher design point of view. This study demonstrates the improvement of the CMP process performance, especially at the wafer edge, from the modification of the floating type polish head. The best WIWNU% from a single air chamber head is about 5.12% at 6-mm edge exclusion (EE). In order to obtain better pad deformation control, the retaining-ring pressure chamber is separated from that of the sub-carrier. The average WIWNU% is about 4% for 3-mm and 5-mm EE from two-pressure-chamber head. Due to the limitation of retaining-ring pressure effect, a third pressure chamber is further added that can be extended the edge control up to 1 inch from the wafer edge. The WIWNU% is about 3.8% at 5-mm edge exclusion with low down forces. The slurry and insert types also show effect on the wafer edge profile. It has been also proven that this three-pressure-chamber head is able to reduce the post-CMP thickness variation from the ILD production wafer, especially at wafer edges. More detailed information and CMP mechanism will be discussed in this paper.

INTRODUCTION

The Chemical Mechanical Planarization (CMP) process, consists of both chemical and mechanical reactions, can generate flat wafer surface due to its an-isotropic nature; therefore, the need of CMP process increases as device feature decreases and integration layer increases.[1] In order to reduce the cost of the chips, the demand of the yield from available dices in each wafer increases also. The reduction of the WIWNU% at small edge exclusion is one of the important factors to achieve higher process yield.

The polish mechanism is that abrasive particles and pad materials contribute to the mechanical removal of wafer surface materials (either dielectric or metal) with the help of slurry chemistries and abrasive surface chemistry.[2,3] Basically the polisher should offer uniform mechanical contact surfaces across both sides of wafer surfaces to obtain homogeneous removal rates. Several researchers have proposed varied removal rate models based on Preston equation. that are all related to down force and linear velocity.[4,5] For polisher with rotational platen, the relative linear velocity (L_V) to the pad surface at varied position (r_w) on wafer surface can be expressed as a function of platen speed(ω_p), head speed(ω_h), and the distance between centers of wafer and platen(r_o).

$$L_V = \omega_p \times r_o + (\omega_p - \omega_h) \times r_w \qquad (1)$$

In order to achieve the same linear velocity on wafer, the matched speed condition ($\omega_p = \omega_h$) should be used. However, the removal rate is still varied (symmetrical or non-symmetrical) even at the matched speed condition due to uneven mechanical contacts. The symmetrical rate variation might be due to the slurry distribution, insert property, pad property variation (includes pad conditioning and pad deformation).[6,7] The un-matched speed condition ($\omega_p \neq \omega_h$) can generate a symmetrical velocity variation across the wafer diameter.(Figure 1) The increase of head speed seems to have more effect on the wafer edge than the reduction of head speed. However, this limited linear speed variation might be useful for metal polish, but it might not be enough for oxide processes. In order to reduce the symmetrical removal rate variation at matched speed condition, it should be solved through the head design.

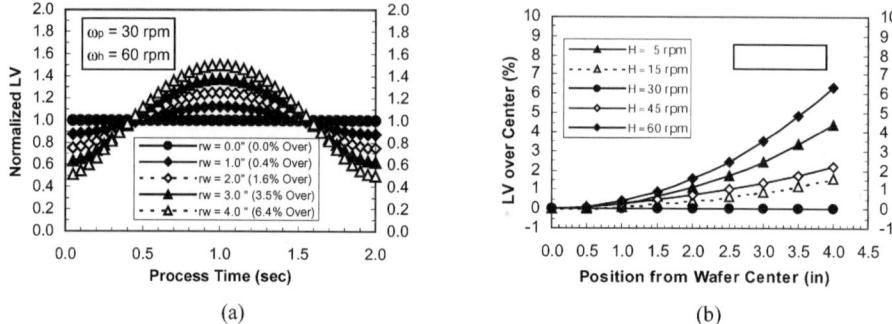

(a)   (b)

Figure 1. Linear relative velocity variation (a) across the wafer with non-matched rotational speed ($\omega_h = 2\,\omega_p$) at varied polish time (b) at varied wafer position with varied non-matched speed at $\omega_p = 30$ rpm.

This study is to improve CMP performance from floating head modification on a polisher with rotational platen. A schematic of floating heads with one-, two-, and three-pressure chambers is depicted in Figure 2. The increase of pressure chambers allows better control on the wafer edge profiles during polish. Three-chamber floating head is also evaluated on patterned oxide wafers. The results show the capability of the third chamber to reduce the thickness variation at the wafer edge after CMP.

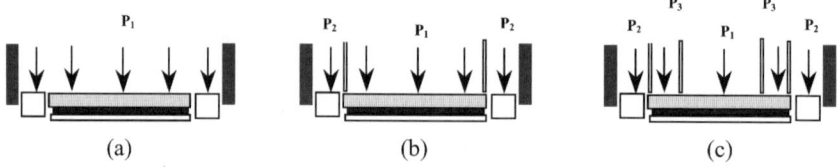

(a)   (b)   (c)

Figure 2. Schematic of floating heads with (a) single, (b) two, and (c) three pressure chambers

EXPERIMENTAL

Commercial available consumables, an IC pad, two inserts (A and B), and two slurries (fumed and colloidal silica) are selected for this test. PE-TEOS blanket wafers are used for the blanket wafer comparison. The polisher is an IP8000 system from Cybeq Nano Technologies. The process parameters are varied for each experiment and will be specified in the result section. The removal rate and non-uniformity (or WIWNU% at 1 σ) are measured with the 49-point contour plots with 5-mm or 6-mm edge exclusion and at an 89-point diameter scan with 2-mm edge exclusion. Patterned oxide wafers from one customer are evaluated for the effect of the third chamber in the three-pressure-chamber floating head.

RESULTS AND DISCUSSION

(1) Comparison of Varied Pressure Chamber Heads from Blanket TEOS Wafers:

Figure 3 shows the process baselines for single- and three-chamber floating heads. Consumables used here are insert A and fumed silica slurry for single-chamber head and are insert B and colloidal silica for three-chamber head. Increasing head speed from the matched speed condition ($\omega_p = \omega_h + \omega_c$) did contribute to the WIWNU% slightly for single-chamber head; however, it didn't correct the slow edge removal process completely. According to Figure 1, there is only 1.5% increase of linear velocity at wafer edge in this process condition. In order to have more control on wafer edge profiles, a second and a third pressure chambers are further introduced. An improved baseline process, removal rate ~ 2000 A/min and WIWNU% ~ 3.8% (1 σ) at 5-mm edge exclusion, for three-chamber head is shown in Figure 3 (b) from the customer.

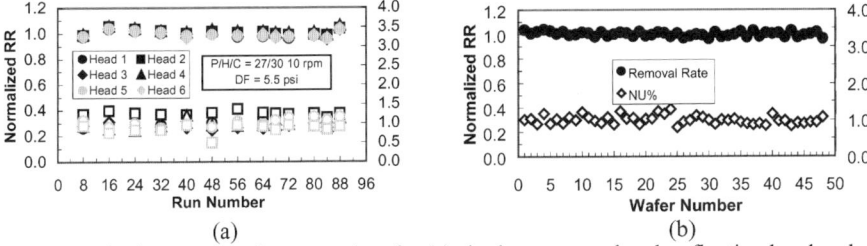

(a)   (b)

Figure 3. Typical process performance data for (a) single pressure chamber floating head and (b) three-pressure chamber floating head with 5.5 psi down forces.

Figure 4(a) shows the effect of the second-pressure chamber, P_2, on the wafer edge profiles. As P_2 increases from 4 to 8 psi, the edge removal characters change from slow to fast removal process. However, the effect of P_2 on the wafer removal profile is only up to about 10 mm from the wafer edge. There are also different wafer edge responses (from wafer edge to about 30 mm further inside) between two types of inserts. Figure 4(b) shows the varied wafer edge profiles generated from two different inserts. Insert A generates better edge profile than Insert B at high down forces, 7 psi. Low down forces, 5.5 psi, and high P_2 give better wafer edge profile for

Insert B. Figure 5 demonstrates the effect of the third-chamber pressure, P_3, on wafer edge profiles at low down forces with colloidal and fumed silica slurries, respectively. The effect of P_3 on wafer edge profile can be extended to about 20 mm inside from the wafer edge. The WIWNU% at 5-mm edge exclusion is improved from the use of P_3 for both slurries. Less third-chamber pressure is required to adjust edge under-polished profile for fumed silica slurry than for the colloidal silica. It only requires 4 psi of P_3 to adjust the wafer edge profile for the fumed silica case. It is believed the morphology and the structure of slurry abrasive particles might play an important role for this phenomenon. The slurry transportation efficiency for both types of slurries might be another factor for this difference.

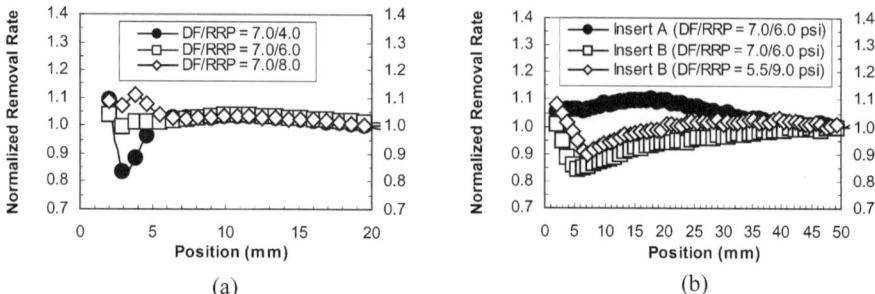

(a)   (b)

Figure 4 Effect of (a) P_2 pressure and (b) varied inserts on the wafer edge profile.

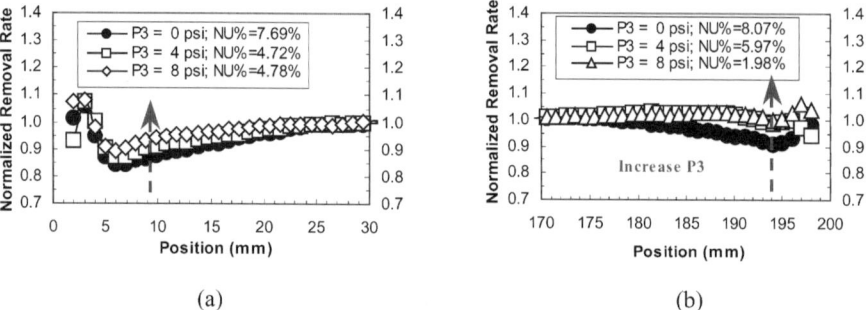

(a)   (b)

Figure 5 Effect of P_3 pressure on wafer edge profiles with (a) Colloidal (b) Fumed silica slurry.

(2) Patterned TEOS Wafers:

Oxide-patterned wafers were polished with three pressure-chamber floating head at one customer site. Totally, more than 3000 wafers were processed at 5.5 psi down force with insert B and with colloidal slurry. Figure 6 shows the typical pre-CMP oxide thickness profile and site positions that will be evaluated for patterned oxide wafers. Figure 7 shows the post-CMP oxide thickness ranges at those nine sites with varied third-chamber pressures (P_3). The removal profile without the third-chamber pressure follows the pre-CMP oxide thickness profile where the same thickness trend is remained for dies located at center and edge of the wafer after polish,

Figure 7 (a). This is a typical response of even removal rate across wafer for most polishers. As the third-chamber pressure increases, the remained oxide thickness variation is reduced. Therefore, an optimized third chamber pressure is very useful to reduce wafer post-CMP oxide thickness variation, especially for those wafers with non-even pre-CMP oxide thickness profile.

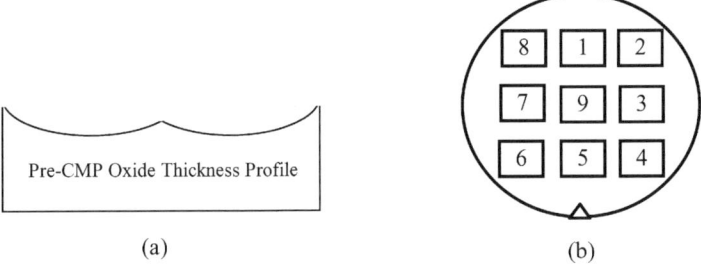

(a)   (b)

Figure 6 (a) Film thickness variation before CMP process and (b) Opti-Probe measurement sites on the patterned wafers.

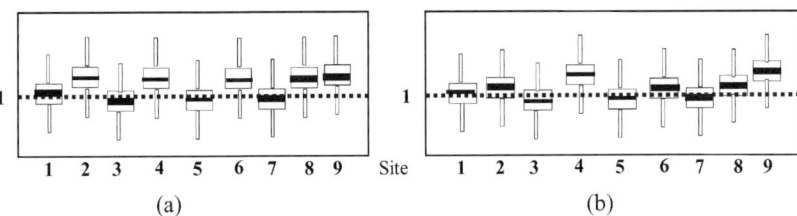

(a)   (b)

Figure 7 (a) Post-CMP thickness variation for 9 sites from 3-chamber head without third chamber pressure, (b) Post-CMP oxide thickness variation for 9 sites from 3-chamber head with the third chamber pressure on, ~ 3 psi.

SUMMARY

The increase or decrease of the head rotational speed in the non-matched speed condition does contribute to a limited symmetrical linear velocity variation across the wafer. The symmetrical removal rate variation at wafer edge is believed from the pad deformation and from the insert. The separation of the second pressure chamber from main sub-carrier chamber allows one to better control pad deformation in CMP process. Therefore, the edge profiles can be altered from edge slow to edge fast process through the control of the second pressure, P_2. However, the effect of the second pressure chamber to the wafer profile is only up to 10 mm from the wafer edge. The use of different insert also contributes to the wafer edge profile variation. The impact of the insert types to the wafer edge profile could be up to 30 mm from the wafer edge. Therefore, a floating head with three pressure-chambers is required to minimize those un-even mechanical portions on wafer polish surface during the CMP process. It has been proven that the third pressure chamber does further improve the wafer edge profile from the use of either fumed or colloidal silica slurries. The third-chamber pressure, P_3, works better to reduce the wafer edge under-polished profiles with fumed silica slurry. It is also demonstrated that the third pressure chamber does reduce the post-CMP oxide thickness variation for the patterned oxide wafers.

REFERENCE

(1)   J. M. Steigerwald, S. P. Murarka, and R. J. Gutmann; *Chemical Mechanical Planarization of Microelectronic Materials*, p22, John Wiley & Sons, Inc., 1997
(2)   F. W. Preston, *J. Soc. Glass Technol*, **11**, p247 (1927)
(3)   L. M. Cook, *J. Non-Cryst. Solids*, **120**, p152 (1990)
(4)   F. Zhang and A. Busnaina; *Electrochem. and Solid-State Letters*, **1** (4), p184 (1998)
(5)   W.T. Tseng, Y. L. Wang, *J. Electrochem. Soc.*, **144** (2), L14 (1997)
(6)   L. M. Cook, J. F. Wang, D. B. James, and A. R. Sethuraman, *Semicond. Int.*, **11**, 141 (1995)
(7)   A. R. Baker, *Proceedings of Electrochem. Soc. Meetings*, **96** (22), p228 (1996)

Mat. Res. Soc. Symp. Vol. 613 © 2000 Materials Research Society

A NEW POLY-Si CMP PROCESS WITH SMALL EROSION FOR ADVANCED TRENCH ISOLATION PROCESS.

Naoto Miyashita*, Shin-ichiro Uekusa**, Takeshi Nishioka*** and Satoko Iwami***
* Dept of Electrical and Electronic Engineering, Meiji Univ., Toshiba Co, Semiconductor
Company 8, Shinsugita-cho, Isogo-ku, Yokohama 235-8522, Yokohama Japan
** Dept of Electrical and Electronic Engineering, Meiji Univ., Kawasaki Japan
*** Mechanical Systems Laboratory, Corporate R&D Center, Toshiba Corporation, 1,
Komukai-Toshiba-cho, Saiwai-ku, Kawasaki Japan

ABSTRACT

Chemical-Mechanical Polishing has been revealed as an attractive technique for poly-Si of trench planalization. Major issue of the process integration is pattern erosion after over polishing. A new process with silica slurry containing organic surfactant is reported in this paper. A patterned wafer after conventional CMP process is eroded by over polishing, however, the new process conducts small erosion for wide trenches. The organic surfactant is well known as a inhibitor for the protection of poly-Si from alkaline, and the new slurry shows a large pH dependency of the viscosity. The experimental work has been focused on the viscosity, and the mechanism of the small erosion is discussed. This new process should be useful for recessing poly-Si by CMP, because it keeps the erosion level very low.

INTRODUCTION

Fine device patterns are formed using CMP techniques in the production of semiconductor devices. Advanced trench isolation technology has been developed and applied to high-speed Bi-CMOS LSI production. Poly-Si CMP technique has made much improvement on the deep trench planarizing process. [1] Major issue of the process integration has been the erosion problem.

In this process, dishing has been caused by over polishing in poly-Si CMP process. During poly-Si CMP process, over-polish is necessary to remove all the residual poly-Si globally for the achievement of high degree of the planarity. Due to the high selectivity of poly-Si removal rate to Oxide by typical poly-Si CMP slurry, over-polish inevitably introduces poly-Si dishing in trench structures. The poly-Si dishing in turn causes oxide erosion and active area's Si3N4 film in the neighboring oxide, attributable to the concentrated cap oxide stress. These stresses introduce many crystal defects. Trench layer should be planarized without dishing on the local scale (distance<700nm) and the global scale(distances>1000nm) in poly-Si CMP process. However it is difficult to reduce the dishing by the conventional poly-Si CMP technique. Therefore we studied a dishing less new CMP method using high viscosity slurry.

EXPERIMENTAL

Figure1 shows a cross sectional image of the trench isolation pattern. The patterned wafers used in this experiment the test structure especially designed for Poly-Si CMP process development. Trench structure with 4500-5000nm of depth is defined and patterned on Si wafer. Before patterning, 100 nm thick thermal oxide and 70 nm thick Si3N4 layers were grown on Si wafers.

LP CVD poly-Si was used for the trench filling. In this trench process, Si3N4 film was used as a stopper for poly-Si CMP and LOCOS musk. Two kinds of films, that is SiO2 and Si3N4 appeared on the wafer surface after poly-Si CMP. Over polish inevitably introduces poly-Si dishing in trench filling poly-Si and Si3N4 erosion on the surface. In this experiment, we paid attention to the viscosity of slurry in order to improve the Si3N4 erosion and poly-Si dishing. Step heights of trench were measured using the probe type apparatus after CMP.

The schematic diagram of the polisher, EBARA EPO-112, used in this study is shown in Fig2. The polisher consists of polishing unit and wafer cleaning unit. The polishing unit is composed of wafer carrier and turntable. The wafer cleaning unit is composed of PVA brush, cloth brush and spin-dry stations [2]. The wafer is held on the wafer carrier and is rotated for CMP. Rodel IC 1000/ SUBA IV stacked polishing pad were applied in this study.

RESULTS AND DISCUSSION

1. Conventional method

Trench filling poly-Si dishing and local Si3N4 erosion is serious issue in poly-Si CMP process. The post-CMP structure is shown schematically in Fig.3. Field Oxide loss is defined as the amount of oxide loss in region with no poly pattern or wide trench pattern. This loss is small typically. Poly-Si dishing and local SiN erosion are then defined as the amount of recess of patterned trench or SiN, relative to filled oxide surface. It is clear that local Si_3N_4 erosion are 1-50nm in our conventional poly-Si CMP process. Poly-Si dishing and local Si_3N_4 erosion associate with stress after trench cap oxidation. In advanced trench process, the wafer after Poly-Si CMP is oxidized as next process step. Therefore, it is necessary to keep the erosion very low. In order to optimize the process, two kinds of slurry mixing CMP process was investigated. The corresponding experiment was produced with two poly-Si CMP slurry. We have measured the profile of total erosion of trench after poly-Si CMP. It is appear that 90-300 nm-erosion is occurred. The viscosity of slurry A is 10-12mPa-s and the pH value of slurry is 10.2-10.5 at the use point of poly-Si CMP.

2. New CMP method

Erosion of trench is serious issue in poly-Si CMP. In order to optimize the process, high viscosity CMP process was investigated. We focused on the viscosity of slurry and increased it in poly-Si CMP process. We have been revealed as an attractive technique for poly-Si of trench planalizing process using two slurry including organic surfactants.

Slurry A is conventional colloidal silica slurry having a 1-5wt% concentration of alkaline abrasive silica content. Slurry B is prepared as the cleaning slurry for post CMP cleaning process.

[4] In the new CMP method, the wafer is polished in two-steps. The first step is the main poly-Si removal step. The second step is dishing less CMP process. In the first polishing step, the slurry A is dropped on to the rotating table. The second step, the slurry B dropped on to the rotating table after optically measured end point signal. At this time the transmittance of slurry were investigated by spectrophotometer. The characteristics of each process step are shown in table 1. Fig.4 shows the relationship of the viscosity of the mixed slurry and pH to the ammonia solvent concentration. The viscosity of the mixed slurry increased rapidly from 0.5% as the alkaline concentration in slurry decreased in our experiment. The pH of mixed slurry decreased rapidly from pH 10.4. In this process step, organic surfactants of slurry B condensed with silica abrasives as core on the table. Fig.5 shows the relationship between pH of slurry B and DI water dilution ratio. The pH value dramatically decreased by using the DI water dilution. In other experiment, the viscosity of slurry B did not increase, because it is not contain silica abrasive.

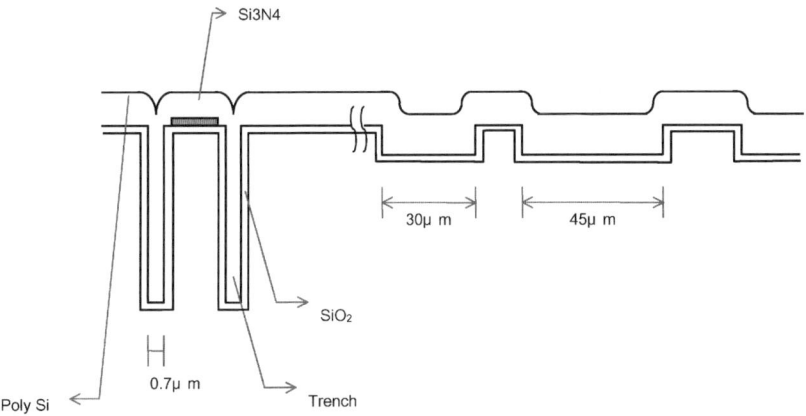

Fig.1   A cross sectional image of the trench test pattern

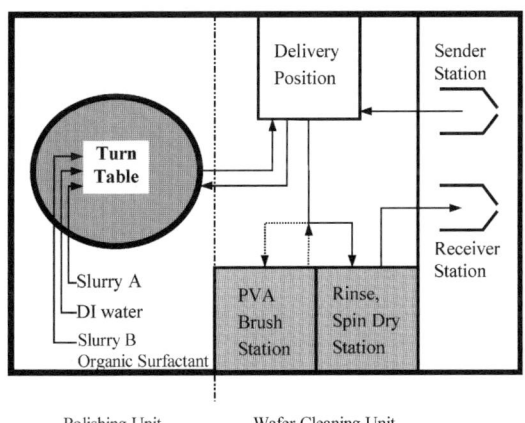

Fig.2   Schematic diagram of the CMP equipment (EPO-112)

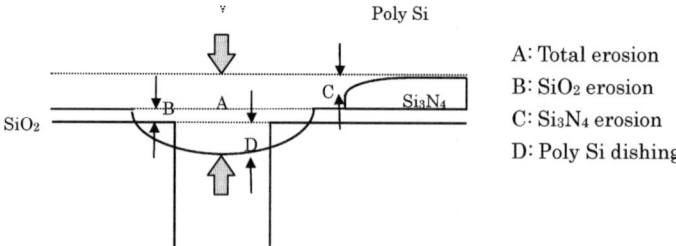

A: Total erosion
B: SiO₂ erosion
C: Si₃N₄ erosion
D: Poly Si dishing

A: Total erosion
B: SiO_2 erosion
C: Si_3N_4 erosion
D: Poly Si dishing

Fig.3 The schematic diagram of trench after CMP

Parameter / Polishing Step	Down Force (Psi)	Table speed (rpm)	Top ring speed (rpm)	Slurry	Polishing rate (nm/min)
1	2~5	100	40	A	800~1000
2	2~5	100	40	A+B	150~170
3	1~2	60	40	B	10~20

Table.1 Estimated characteristics of the process step

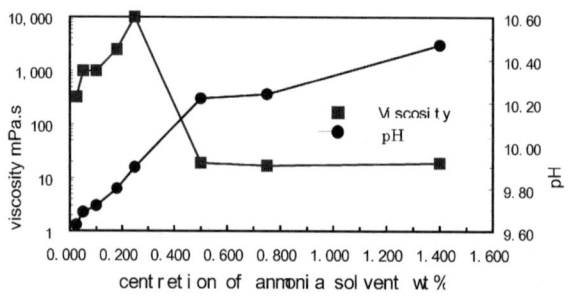

Fig.4 The relation of viscosity of mixed slurry and pH of anemone solvent on the polishing table

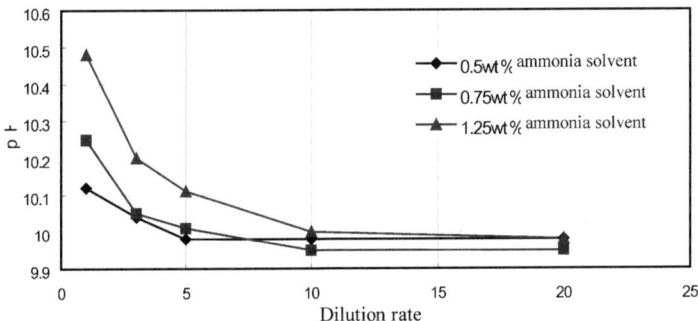

Fig.5 The relation of slurry B pH to DI water dilution ratio.

By using this new CMP method, the erosion of trench on wafer is below 5nm (7nm width) trench structure and below 50nm(30-50 um width) recess. Two possible mechanisms are considered for the excellent planarity. One is that the condensed surfactants make the pad harder, and the other is the increase of the hydrodynamic pressure of the slurry. Nishioka, et.al. proposed a model on the hydrodynamic effects of pad surface roughness, and reported the effective hydrodynamic pressure occurs with more than 100 mPa-s viscosity of slurry [5]. Farther investigation should be necessary to make the mechanism clear.

CONCLUSIONS

From these results, we conclude that:

(1) The poly-Si of trench after poly-Si CMP process were eroded by over in the case of conventional method.

(2) This new CMP method with the organic surfactants of slurry could reduce the erosion of the poly-Si of trench on the wafer below 5nm and below 50nm(30-50 um width) recess. It is applicable to CMP planarization process for VLSI.

(3) The mechanisms of the excellent planarity are considered as the effect of the condensation of the surfactants and the increase of the viscosity of the slurry.

ACKNOWLEDGMENT

The authors would like to thank Mr. Takayasu and Mr. Minami who has made a great effect to do these experiments. And authors gratefully acknowledge the personnel in the Silicon Facility at TOSHIBA Micro electronics center for processing the wafers. Special thanks go to S.Kikuchi and M.Terasaki for the lithography, K.Doi for the trench fill, Y.Otani for the trench etching, and K.Iwade for the line process.

REFERENCES

[1] S. A. Abbas : "Silicon on poly silicon with deep dielectric isolation", Proc. of IBM' Technical Disclosure Bulletin Vol. No.7 Dec. 1997 P.2754
[2] D. L. Hetherington : "The effects of double-sided scrubbing on removal of particles and metal contamination from chemical-mechanical-polished wafers", Proc. of DUMIC Conference 1995 ISMIC-101D/95/0156
[3] N. Miyashita, Y. Minami, I. Katakabe, J. Takayasu, M. Abe, T.Izumi: "Characterization of new post CMP cleaning method for trench isolation process" Proc. of Proceedings of 14th International Vacuum Congress. (1999)P.71
[4] N.Miyashita, Y.mase, J.Takayasu, Y.Minami, M.Kodera, M.Abe, and T.Izumi: "Mechanism of a new post CMP cleaning for trench isolation process." Proc. of Proceedings of MRS.Vol.566 (1999)P. 253
[5] T.Nishioka, K.Sekine and Y.Tateyama: "Modeling on hydrodynamic effects of pad surface roughness in CMP process" Proc. IEEE 1999 Int. Interconnect Tech. Conf. (1999) p.89

Mat. Res. Soc. Symp. Vol. 613 © 2000 Materials Research Society

The characteristics of the Electrolyzed D.I.water with chemicals and the outline of the supply system

Mitsuhiko Shirakashi*, Kenya Itoh*, Ichiro Katakabe*, Mamabu Tsujimura*,
Takayuki Saitoh**, Kaoru Yamada**,
Naoto Miyashita***, Masako Kodera***, Yoshitaka Matsui***

    * Precision Machinery Group, Ebara Corporation
      4-2-1,Honfujisawa, Fujisawa-shi, Kanagawa 251-8502, Japan
   ** Center of Technology Development, Ebara Research CO., Ltd.
      4-2-1,Honfujisawa, Fujisawa-shi, Kanagawa 251-8502, Japan
*** Manufacturing Engineering Center, Toshiba Corporation Semiconductor Company
      8,Shinsugita-cho, Isogo-ku, Yokohama, Kanagawa 235-8522, Japan

ABSTRACT

In semiconductor device production, wafers are treated through many cleaning processes. Usually, several chemicals are used so as to match for several purposes like RCA cleaning method. As wafer size becomes larger, large amount of cleaning chemicals usage and waste are necessary, which becomes now a big problem. Considering the above, we developed the Electrolyzed D.I.water with chemicals supply system in order to minimize running cost of chemicals and waste treatment. It is feature that; 1) this system can generate the anode water of the acidity/high ORP (Oxidation-Reduction Potential) and the cathode water of the alkalinity/low ORP by electrolyzing D.I.water adding with a small quantity of chemicals; 2) this system can generate anode water and cathode water at the same time; 3) if necessary, the anode water can be diluted with D.I.water at the optional density, and it is possible with the cathode water that hydrogen peroxide is added. The anode water which shows acidity/high ORP has the effect of removing metal and organic contamination, and the cathode water which shows alkalinity/low ORP has the effect of particle removal.
In this report, we explain the outline of this system and the basic characteristic of the Acid and Alkaline water made with this system and its performances.

INTRODUCTION

RCA cleaning process using hydrochloric acid, ammonia, sulfuric acid, hydrofluoric acid etc. has been well adopted so far. However, in accordance with larger wafer size, it is very important to minimize chemical cost and waste treatment cost. Through above situation, in 90's, new idea of chemical free or less has been highlighted as functioned water like gas-dissolved water. The Electrolyzed D.I.water with chemicals is developed in accordance with above requirement.

The cleaning mechanisms are defined as follows.
(1) Physical effects of removal foreign materials from cleaning surface using blush or jet.
(2) Electrical potential is controlled by pH change.
(3) Etch off of cleaning surface by chemical
(4) Remove foreign materials by reacting with chemicals

The anode water is effective for (4) and the cathode water is effective for (2) & (3).

EXPERIMENTAL

First of all, we explain the principle of generating the Electrolyzed D.I.water with chemicals on Figure 1.
This cell has an ion exchange membrane, anode and cathode electrode and electrolyzes D.I. water adding chemicals. HCl is introduced to anode side and HCl/NH$_4$OH are for cathode side. Reactions of anode and cathode sides are as follows.

Anode side: $2H_2O \rightarrow 4H^+ + O_2 + 4e^-$
$2Cl^- \rightarrow Cl_2 + 2e^-$
$Cl_2 + H_2O \rightarrow HCl + HClO$
Cathode side: $2H_2O + 2 e^- \rightarrow 2OH^- + H_2$

Cl_2 and HClO are generated in anode side. Cl_2 and HClO have oxidization performance.
This anode water (named Acid water) is effective to remove metal contamination by combination of HCl, HClO and Cl_2. It stands at 0-2 of pH and more than 1000mV(vs. Ag/AgCl electrode) of ORP.
In cathode side, this cathode water (named Alkaline water) is effective to remove particles. It stands at 9-11 of pH and less than –500mV(vs. Ag/AgCl electrode) of ORP.

Usually oxidized metal has been used for electrode in electrolyzing cells. In our new developed electrolyzing tool, special carbon material is adopted in order to avoid corrosion and metal contamination. Carbon crack is also overcome by special surface treatment. Metal contamination in these waters is less than 100 ppt.

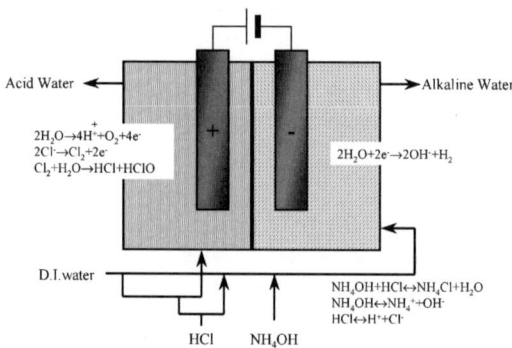

Figure 1 : Principle of generating the Electrolyzed D.I.water with chemicals

Figure 2 shows flow of the Electrolyzed D.I.water with chemicals supply system.
The unit has one electrolyzing cell, which generate Acid and Alkaline water at the same time.
It is possible that acid water dilute in any ratio with D.I.water and hydrogen peroxide is added to alkaline water.
Figure 3 shows appearance of the Electrolyzed D.I.water with chemicals

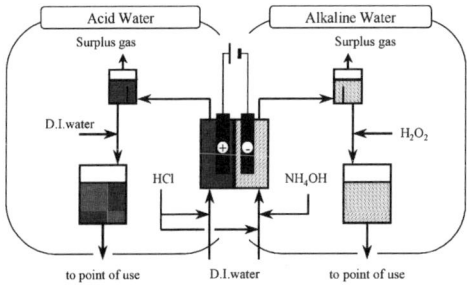

Figure 2 : Flow of the Electrolyzed D.I.water with chemicals supply system

Figure 3 :
Appearance of
the Electrolyzed D.I.water with chemicals
supply unit

DISCUSSION

At first, we explain performance of metal contamination removal using Acid water.
Test procedure is shown on Figure 4 and result on Figure 5.
Bare-Si wafer, on which Fe, Ni and Cu are coated, is used for this test. Procedure is this wafer is rinsed by D.I.water and then by Acid water during spinning. The evaluation is made by TRXRF (Total Reflection X-ray Fluorescence).

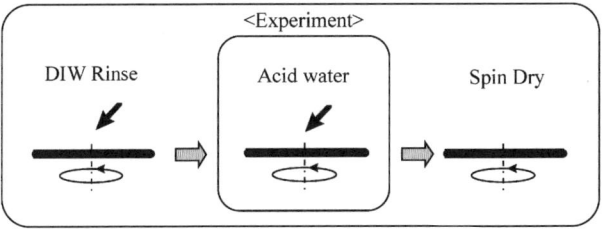

Figure 4 : Procedure of experiment

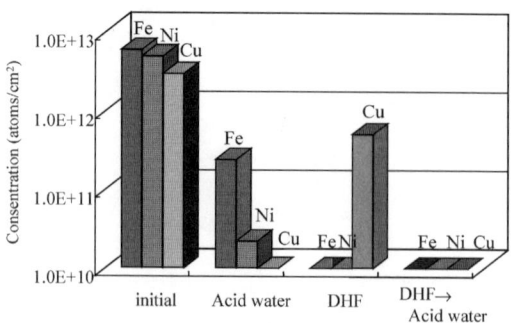

Figure 5 : Result of metal contamination removal

Only Acid water and only DHF (diluted HF) aren't almighty, but if we choice the effective combination, it is very effective. Figure 5 shows that metal contamination is reduced from 1×10^{12} level to ND level (under 1×10^{10}) by Acid water treatment after DHF.

Next is the result of particle removal by Alkaline water.
Figure-6 shows test procedure.
Bare Si wafer, on which Al_2O_3 slurry is coated, is used for this test.
Procedure is that wafer is rinsed by D.I.water and then cleaned by Alkaline water with ultrasonic nozzle during spinning. The evaluation is made by SP-1 (KLA-Tencor)

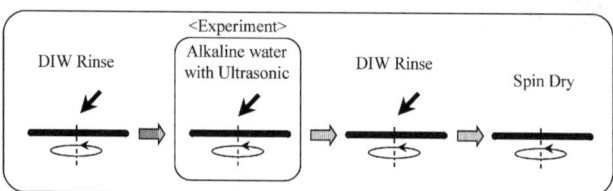

Figure 6 : Procedure of experiment

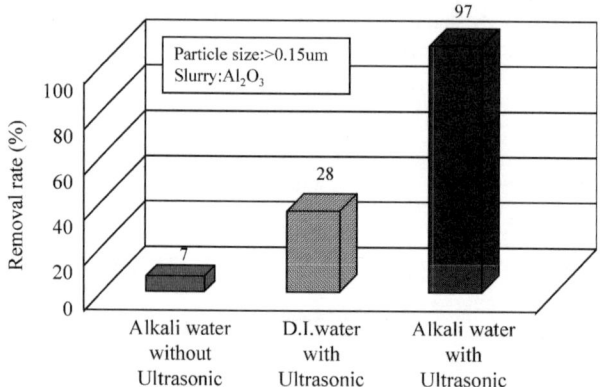

Figure 7 : Result of particle removal (Wf:SiO_2)

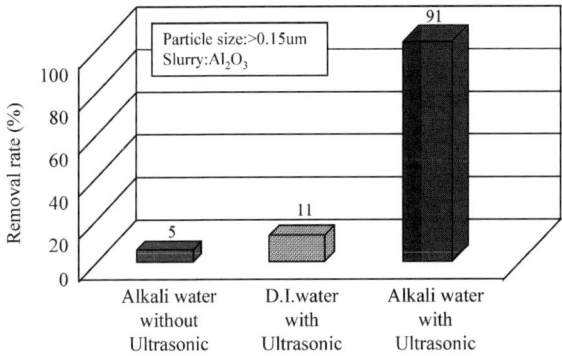

Figure 8 : Result of particle removal (Wf:Bare-Si)

Figure-7 shows test result of SiO$_2$ wafer.
Removal efficiencies are only 7% by pouring Alkaline water, 28% by D.I.water with ultrasonic nozzle and 97% is obtained by Alkaline water with ultrasonic nozzle.
This result shows combination of Alkaline water and ultrasonic nozzle is very effective.

Figure-8 shows test result of bare-Si wafer.
Removal efficiency is more than 90% by Alkaline water with ultrasonic nozzle in spite that case of bare-Si and Al$_2$O$_3$ is much more difficult than case of SiO$_2$ and Al$_2$O$_3$ to remove particle contamination.

From these results, it is concluded that Alkaline water is effective to remove particle contamination.

CONCLUSION

We acquired following knowledge from above experimental.
(1) Acid water is effective to remove metal contamination as result shows to reduce from 1×10^{12} level to ND level.
(2) Alkaline water with ultrasonic nozzle is effective to remove particle contamination.
It is confirmed that the Electrolyzed D.I.water with chemicals is effective as general chemical cleaning method.

We developed ultra pure cleaning liquid that don't be generated by oxidized metal electrodes and shows better performance than usual chemical cleaning method in spite of adding a bit of chemicals.

REFERENCES

M.Kashiwagi, A.Hattori: "Cleaning technology of wafer surface", Realize INC.

Poster Session:
CMP Consumables

An Image Analysis Technique For Assessing Particle Size And Agglomeration Tendency Of Slurries

Susan R. Machinski[1,2], Kathleen A. Richardson[1,2], and Aristide Dogariu[2]
[1] Department of Chemistry
[2] School of Optics, CREOL
University of Central Florida, 4000 Central Florida Boulevard
Orlando, FL 32816

ABSTRACT

Understanding the time-dependent behavior and size of particulate systems, specifically, particles in abrasive slurries, is a key part of chemical mechanical polishing (CMP). Microscopy and image analysis enables both qualitative and quantitative analysis of particle interaction behavior in diluted samples of such particulate suspensions. A technique using microscopy and image analysis has been developed specific to the analysis of suspensions of, micron size or larger, abrasive particle slurry systems. This technique has been applied to aluminum oxide (Al_2O_3) slurries and measurements of particle size and stability have been obtained with good accuracy and precision.

INTRODUCTION

A wide variety of particulate materials have been used as the mechanical abrasive chosen to grind (coarse removal) or polish (fine) solid surfaces. These include Al_2O_3, SiO_2, SiC, and Si_3N_4. An important issue in understanding the chemical and mechanical processes in optical glass polishing systems is the nature of the abrasive slurries. The nature of the crystalline particle (single versus polycrystalline), particle size, and concentration of the abrasive in the slurry, as well as the stability of the system, all have a direct impact on the polishing mechanism and eventually on the resulting quality of the finished workpiece [1]. Currently, the abrasive slurries used in semiconductor wafer processing are used once, and not recycled in the process. By understanding the particle sizing and the stability of the dispersed solid-liquid system, such as those used in the CMP industry, optimal end use shelf-life for the abrasive slurries may be determined. Thus, the analysis of particle sizes in the slurries over a period of time can provide a relationship between the abrasive slurry particle and the point of its resultant degradation in the polishing process.

Abrasive slurry particles tend to form aggregates which then may agglomerate into large clusters. Base-stabilized (BS) and acid-stabilized (AS) Al_2O_3 slurry systems remain dispersed over short working times but, have been reported to agglomerate over time [1]. Large abrasive clusters can lead to scratching, surface defects, and if not completely removed following polishing, surface contamination. It is this agglomeration behavior that must be controlled. Understanding the agglomeration process is further complicated by the interaction of particles with the aqueous slurry solution. There is believed to be a direct correlation between micro scratching and the presence of larger particles [2].

Most particle sizing algorithms are calculated by the spherical equivalence principle [3]. However, if the shape of the particle is non-spherical, different sizes will be measured by different methods [3]. The methods used for particle sizing are dependent of the type of sample under analysis. Each method also possesses its own set of limitations. For example, a widely used technique for sizing the abrasive particles is laser light scattering. The theory of the technique centers on the premise that different sized particles will scatter light at different angles. By measuring the intensity of light at different angles a particle size distribution can be calculated. However, there are a variety of algorithms used to calculate the distribution, which can result in different size distributions for the same sample [4]. It is also assumed that the particles are spherical. This research used the image analysis software program, OPTIMAS 6.0, which uses a best fit polygon to measure areas of objects. This allows for a more precise measurement of particle size. OPTIMAS is a software program which allows for image processing and measurement of features within the image [5].

This paper reports on the use of optical microscopy and image analysis in the characterization of BS Al_2O_3 slurries. The work resulted from the investigation of Lauwidjaja which compared stability in alumina slurry systems using electrochemical and optical (enhanced backscattering (EBS)) techniques [1]. As electrochemical techniques required dilution of the slurry system under study (>400X) and EBS does not, a quantitative measure of particle size and distribution as a function of slurry concentration, as a function of slurry age, was sought. The use of optical microscopy allows for visual inspection and characterization of the particles as well as quantitative image analysis to obtain individual particle size distribution as a function of time and chemical environment. Hence, the motivation for this study was to use optical microscopy with image analysis to ascertain quantification of particle size, its variation, and to see if a) a change in distribution could be seen in a commercialized "stabilized" slurry, and b) if such a distribution change could be used as a gauge of slurry stability.

EXPERIMENTAL PROCEDURE

For this study a particle was defined as the smallest unit of the slurry system and an agglomerated particle was many smaller particles gathered closely together and in contact.

The pH of the slurry samples was measured with an Accumet pH meter [6] equipped with a rugged tip probe [7]. pH of the slurry fluid to be examined was measured prior to preparation of the sample on a glass microscope slide.

The instrumentation employed for assessing the particle size and agglomeration tendency of the Al_2O_3 slurry was a Nikon Labophot-2A [8] optical microscope equipped with a Hitachi KPD50 1/2" color CCD [9] with high-resolution camera. A transmitted objective (40X) corrected for coverslips of 0.17 μm thickness was used for photo microscopy. An Intel 200 MHz Pentium processor computer with MMX technology and OPTIMAS 6.0, a windows-based image analysis software was employed.

Size calibration samples were used as an absolute size standard reference to verify and ensure the accuracy of the quantitative measurements obtained from the OPTIMAS program and to assess modification to acquired images imparted by the image analysis system variables (filters, thresholds, intensity, etc.). The calibration samples were 3.7 μm diameter polymer latex spheres in an aqueous suspension (Duke Scientific Corporation) [10]. They were air dried to produce dry solid spheres prior to being used as calibrated standards in this study. As no other standards were available for this study, no systematic evaluation of standard size versus imaged size was carried out.

Two experimental slurry samples were analyzed; BS Al_2O_3 slurry and AS Al_2O_3 slurry, both produced by Norton Corporation [11]. This paper only reports results obtained on the BS Al_2O_3. The BS Al_2O_3 as-received was 24% by weight solid. It was described by the manufacturer to contain sodium polyacrylate as a dispersant along with other additives that were unknown due to the proprietary nature of the slurry. The pH of the as-received BS Al_2O_3, was reported by the manufacturer to be between 9.5 and 10. Five different aqueous dilutions of the BS Al_2O_3 were prepared using deionized water. The corresponding sample concentrations of slurry:deionized water were 0.24, 2.0, 7.2, 12.0, 16.8 % by weight. The prepared diluted sample concentrations remained static (i.e. unagitated) for the duration of the experiment in sealed glass containers. The samples aged in a temperature-controlled room in which the temperature was monitored throughout the study (and at the onset of each day's sample measurement) to be stable at 23.5 °C.

The BS Al_2O_3 prepared samples were aged at their prepared concentrations. The initial slurry mixture dilutions were concentrated and too opaque for viewing individual particles. Hence, a 1:200 dilution for all slurry samples was chosen after experimenting with various dilution levels under the microscope. Dilution was carried out immediately prior to microscopic analysis, however, aging between measurements was carried out in the more concentrated form.

Initial measurements and microscopic images were taken 2 hours (Day 0) after preparing the slurry dilutions. Subsequent measurements and microscopic images of the aged samples were again taken 24 hours later (Day 1) and thereafter every two days at approximately the same time of day for a period of two weeks. The entire sample on the slide was first viewed at a 10X magnification for overall representation of the sample. Then, utilizing the 40X transmitted objective, photos of three representative areas from different areas of the slide where taken with the camera and saved in the OPTIMAS program.

After all data was collected for the two week period, particle sizing image analysis began. First, an OPTIMAS procedure was developed for the standard and applied to all slurry microphotographs. Second, the OPTIMAS data was exported to EXCEL for further analysis and graphical representation.

DISCUSSION

I.   Reference Samples and System Calibration

In trying to find the most precise procedure for particle area quantification of it was noted that applying different filters and a gray morphology had an effect on the particle area measurement quantified by OPTIMAS. This was supported by the calibration standard data in Table I. A distortion of the image was seen in the microphotographs of the individual image analysis procedural steps (not shown here) regardless of the enhancement. It is interesting to note that the data in Table I indicated that application of the gray scale alone resulted in the highest deviation ($+1.07\mu m$) from the true reference diameter ($3.7\mu m$). However, lower deviations from the standard reference particle resulted when the gray scale was combined with a filter application. This was believed to be due to the blurring and fuzziness of the edges of the particles due to pixel display and inherent noise in the image on the computer screen prior to a filter being applied. It can only be concluded that the filters were indeed doing their job of reducing noise and increasing resolution. The best average diameter measurement that could be obtained with OPTIMAS for the calibration standard particles was $4.30 \mu m$ using a gray scale and a median 3x3 filter. This resulted in a diameter measurement error of $+ 0.60 \mu m$. Being the best possible procedure for optimizing the calibrated standard particles for analysis, this procedure was applied to all microphotographs of BS Al_2O_3 slurry particles. It should be reiterated that all subsequent measurements using this "optimized" procedure would be subjected to a similar $+0.6 \mu m$.

The images of the "best" procedural steps as applied to the slurry samples (not showed here) also showed visually, little distortion of the particles. There were some particles not identified by the threshold. Elimination of some small particles was a compensation for over sizing the particles. A maximum of approximately 27 % of particles was calculated to not be taken into consideration in the measurements. These particles were estimated to be under 0.1 μm in area. These particles appeared much lighter in color than the particles which the threshold determined. This may very well be one of the disadvantages of image analysis. As pixels are enhanced through the image manipulation using intensity adjustment and filters, there occurs inherently on the monitor image a variation of the image's gray background resembling shadows.

The determination of the threshold may provide for the greatest source of error in OPTIMAS image analysis. This step was completely subjective to the operator's perception of particle size. Consistency as to the determination of the threshold, i.e. just under, just over, etc. must be applied to each picture. The threshold can be manipulated after initially applied to make adjustments as needed. Due to the window nature of the OPTIMAS program, a comparison can be made simultaneously between the threshold and non-threshold image thereby allowing the operator to obtain the most accurate threshold.

Table I. Average area and Average Diameter for 3.7 μ Diameter (Area = 10.75 μm^2) Calibration Standard using Different Enhancements.

Enhancement	Average Area (μm^2)	Average Diameter (μm)
Gray scale only	17.87	4.77
Gray scale + gray morphology	16.40	4.53
Gray scale + median 3x3	14.52	4.30
Gray scale + gray morphology + gaussian 3x3	16.33	4.56

II. Slurry Stability Analysis

Data obtained on the BS Al_2O_3 during this study are presented in Table II. The **average area** obtained showed a significant decrease in area for all concentrations with aging from Day 0 to Day 13. It is interesting to note two measurements, pH and **most frequent area**. With the exception of the 0.24% by wt. concentration, the suspension pH showed an increase with increasing concentration for each Day 0 and Day 13 but each concentration remained quite stable from Day 0 to Day 13. The 0.24% by wt. concentration pH increased 0.67 pH units over the thirteen day period. The remaining concentrations, 2.0, 7.2, 12.0, and 16.8 % by wt., were measured to decrease minimally (0.06, 0.05, 0.00, 0.17, pH units respectively). It is noted that for the more concentrated samples (7.2, 12.0, and 16.8 % by wt.) the pH remained at approximately the pH of the as received BS Al_2O_3 concentration (9.5 - 10.0 and measured at 9.38). The 0.24 and 2.0 % by wt. concentration samples were measured to be lower than that of the as-received BS Al_2O_3 but remained stable over the aging period. The pH results indicated that deionized water (pH ~5) does not effect the pH stabilization of the slurry except at extremely low concentrations of under 2.0% wt. This was expected since deionized water imparts no charged species to the suspension which, if present, could modify particle surface charge and agglomeration tendencies. The measured pH of the deionized water used was initially 5.07. Thus it was also expected that at very low concentration levels of BS Al_2O_3 and high concentrations of deionized water a shift in pH must occur to stabilize at some pH unit in-between the alumina and deionized water.

The **most frequent area** for the particles was between 2-3 μm^2 for all concentrations with the exception of 2.0 % by wt. concentration on Day 0. This was interesting because the average agglomerate area is much higher (from 10.8μm^2 to 60.14μm^2) again with the exception of 3.4μm^2 for the 2.0 % by wt. concentration. Comparing images of sample concentrations from Day 0 through Day 13 a definite decrease in size was visually observed (not shown here). The microphotographs in Figure 1 illustrate this observation for the 12.0 % by wt. concentration and are representative of the other concentrations. The image for Day 0 (initial prepared diluted slurry) showed the slurry particles were large and agglomerated. The image for Day 13 (diluted slurry after aging for thirteen days) showed the particles to be smaller and dispersed. This relationship is also shown in Table II as calculated from OPTIMAS.

The **average area** decreased for all aged concentration samples from Day 0 to Day 13. A possible explanation for the particle size decrease was that on Day 0 a disruption in the slurry's equilibrium occurred due to introduction of water molecules. The polyacrylate dispersant was also believed to be disrupted in its function. The alumina aggregates are formed as a result of this disruption to the system during this initial period. By Day 13, equilibrium was restored between the dispersant and the alumina. This result as observed using microscopy was consistent with results of Lauwidjaja using other techniques for particle surface chemical analysis such as zeta potential [1].

Table II. Measurement Data Obtained Using OPTIMAS for BS Al_2O_3

Day 0							
	Conc.	# Particles Meas.	Avg. Area (μm^2)	Median Area (μm^2)	Most Freq. Area (μm^2)	R.M.S.	pH
	0.24%	756	20.09	6.39	2-3	3.48	6.17
	2.0%	432	25.76	6.1	1-2	2.3	8.23
	7.2%	137	17.41	4.29	2-3	5.19	9.44
	12.0%	170	45.14	4.39	2-3	.627	9.58
	16.8%	192	60.14	6.29	2-3	.370	9.63
Day 13							
	0.24%	310	14.35	4.20	2-3	5.96	6.84
	2.0%	376	3.40	2.57	2-3	4.25	8.17
	7.2%	685	10.82	6.30	2-3	5.83	9.39
	12.0%	441	22.19	4.58	2-3	6.12	9.58
	16.8%	347	22.96	5.05	2-3	4.81	9.46

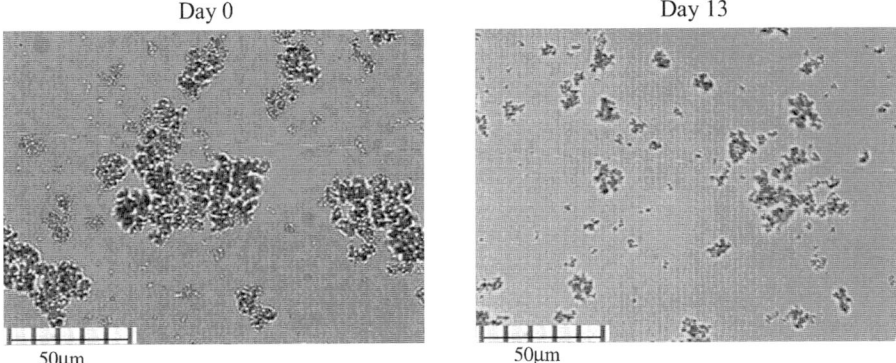

Day 0   Day 13

50μm   50μm

Figure 1. 12.0 % by wt. BS Al₂O₃ compared at Day 0 (after initial sample preparation) and Day 13 (aged), diluted to 1:200, viewed with a 40X transmitted objective.

The total number of particles measured in the representative regions of interest (ROI's) was found to be inversely proportional to the sample concentration at Day 0. This relationship was not observed at Day 13. However, from Day 0 to Day 13 an increase in the number of particles measured was obtained for the three higher concentrations and a decrease was observed for the two lower concentrations. Again this may be explained as a function of the disruption to equilibrium due to deionized water as described above.

The BS Al₂O₃ was shown to have some concentration effects. Graphical comparisons of particle area sizes' distribution for each BS Al₂O₃ sample concentration at Day 0 and Day 13 (not shown here) showed these effects were most pronounced for Day 13. The 2.0 % by wt. BS Al₂O₃ showed an approximate doubling in frequency for particles between 2 and 6 μm². This trend diminished below 6 μm² and fell to nearly zero particles greater than 10 μm². For the remaining concentrations particles were found shown to be bimodal in distribution for both Day 0 and Day 13. This bimodal trend, however, was decreased in Day 13. The frequency of smaller particles showed an increase with Day 13. The trend of an increase in smaller particles was verified again in the frequency vs. standard deviation (S.D.) charts for the all concentrations of BS Al₂O₃. Figure 2 illustrates this trend as was seen in the 12.0 % by wt. concentration. This trend was seen in the microscopic images. However, close examination of the concentration vs. day charts seem to show that the particles did not significantly decrease over time but that the smaller particles agglomerated to give a continuous distribution curve.

CONCLUSION

An optimized procedure for acquiring and analyzing images of particulate systems was determined. This procedure was successfully applied to the analysis of a BS Al₂O₃ slurry particulate system. Optical microscopy visually showed the slurry particles to decrease with aging. This was confirmed by image analysis which measured the **average area** to decrease. However, no appreciable change in the **most frequent area** occurred with aging. Graphical representation of the image analysis data showed a broadening in distribution of the particle size with aging. The graphical representations seemed to indicate that the smaller particles agglomerated rather than the larger particles becoming non-agglomerated. These results, indicating destabilization of Al₂O₃ slurry systems with aging, agreed with others using zeta potential (dilute concentrations) and EBS (more concentrated samples) [1]. In that EBS used a simplistic solid, spherical particle method to calculate average particle size, the particle morphology and non-spherical particle size calculation obtained with microscopy and image analysis can provide a more refined fractal aggregation theory for EBS analysis.

Microscopy and image analysis using this procedure can provide the ability to track, qualitatively and quantitatively, agglomeration behavior of particulate systems over a period time. Further work is needed to identify complementary analysis tools which when used in conjunction with microscopy and image analysis (i.e., dynamic EBS) can yield a direct measure of slurry stability. Ideally, the probing of the dynamic processes as a function of real time is possible.

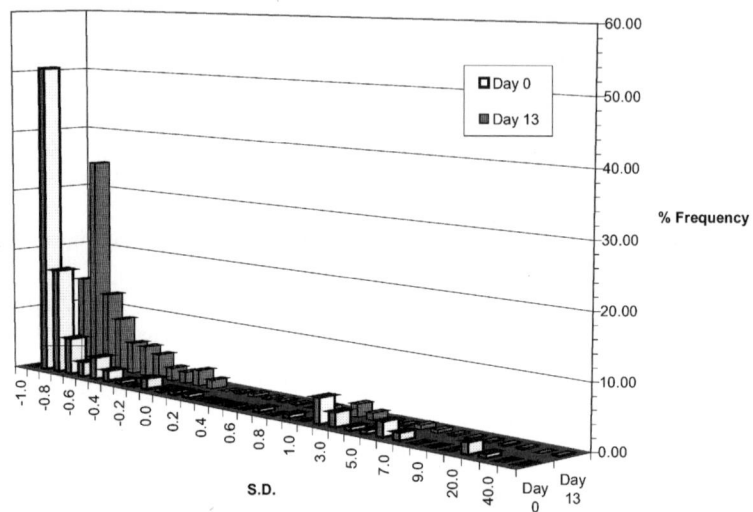

Figure 2. Percent frequency for standard deviation from average area particle size of 45.14 μm^2 for 12.0 % by wt. BS Al_2O_3.

REFERENCES

1. Lauwidjaja, M., Characterization of the Stability of Aluminum Oxide (Al_2O_3) Slurries for Optical Polishing, M.S. thesis, Univ. of Central FL, 1998
2. Pohl, M.C., Griffiths, D.A., J. Elect.Mater., **25**, 1612 (1996)
3. Bernhardt, C., *Particle Size Analysis; Classification and Sedimentation Methods*, (Chapman & Hall: London, 1994), p. 4
4. Pohl, M.C., Griffiths, D.A., J. Elect. Mater., **25**, 1612 (1996)
5. Media Cybernetics, 8484 Georgia Avenue, Suite 200, Silver Spring, MD 20910
6. Model 50, Denver Instr. Co., 6542 Fig St., Arvada, CO 80004
7. Catalog #13620181, Fisher Scientific
8. Southern Micro Instrumentations, 8250 Exchange Dr., Suite 110, Orlando, FL 32809
9. Model KP-D5OU, Nikon, 6420 Dobbin Rd., Suite E, Columbia, MD 21045
10. Catalog #5370A, Lot#17230, MSDS, Duke Scientific Corp., 2463 Faber Place, Paloi, CA 94303
11. BS Al_2O_3 Lot #910407, Product Code 9220, Norton Co. Materials, 1 New Bond Street, Worcester, MA 01615

Mat. Res. Soc. Symp. Vol. 613 © 2000 Materials Research Society

Performance of Polishing Slurries containing Silica Particles grown by Sol-Gel Method

Sun Hyuk Bae, Jae-Hyun So, Seung-Man Yang and Do Hyun Kim
Department of Chemical Engineering
Korea Advanced Institute of Science and Technology
373-1, Kusong-dong, Yusung-gu,Taejon, 305-701, Korea

ABSTRACT

Silica slurry used as abrasives in wafer polishing process is made by dispersing silica particles in an alkali solution. Since commercially available colloidal or fumed silica particles need some modifications to be directly used as abrasive slurry due to their small sizes, irregular shapes or broad size distribution, we have prepared silica abrasives by particle growth of fumed silica or colloidal silica as seeds by sol-gel method. Silica slurries prepared by this step-wise growth from commercial seeds were tested using one-armed polisher for the comparison with commercial slurries and showed the performance comparable to commercial slurries. Microstructures of polishing slurries were investigated using transmission electron microscopy and ARES rheometer. From the result, stability of the slurry was found to be more important than the primary particle sizes for the polishing performance.

INTRODUCTION

Semiconductor industry requires silicon wafers with extremely tight specifications with respect to flatness and surface uniformity. Thickness variations in the sub-μm range as well as RMS (root mean square) roughness values in the sub-Å regime are prerequisite to efficient manufacturing of advanced integrated circuits. Wafer polishing process consists of stock removal polishing and final polishing step. Wafer surface is effectively removed in stock removal polishing step and final polishing determines the flatness and surface smoothness of the silicon wafers in the wafer manufacturing process. In the polishing process, silicon wafer is polished by chemical and mechanical action between silica particles in the slurry and silicon wafer surface. Polishing slurry is prepared by dispersing silica particles of nm size in a alkaline solution like ammonium hydroxide solution. Among several additives, surfactants are used for the stabilizing of colloidal slurry since surfactants with amphiphilic radicals provide steric or electrostatic hindrance between particles in the slurry.

Sol-gel method has long been used to prepare silica particles. Stöber *et al.* synthesized monodispersed particles by hydrolysis and condensation of silicon alkoxide using ammonia as catalyst [1]. However, the slurry containing silica particles *per se* prepared by Stöber method has too low solid content to be used as a polishing slurry. Bogush *et al.* used silica particles by Stöber method as seeds to increase the size of particles and developed the relationship between particle size and concentrations of TEOS (tetraethylorthosilicate), ammonia and water [2]. More concentrated particle dispersion with narrow size distribution using Ludox AS-40® (trademark of colloidal silica of Ludox Corp.) as seeds was reported [3]. Commercially available silica particles such as Ludox® or Aerosil® (trademark of fumed silica of Degussa Corp.) have some disadvantages to be directly used for abrasive slurry because of small size of silica particles, broad size distribution or irregular shapes.

In this study, silica particles were grown by hydrolysis and condensation of TEOS using fumed or colloidal silica as seeds in order to provide proper size of particles for polishing slurry. Then, alcohol-base solvent in the slurry containing grown silica particles is substituted by water solvent through vacuum evaporation. Additives are added to the slurry with water solvent to promote the polishing rate and to enhance the stability of particle dispersion taking advantage of electrostatic hindering action. Polishing test was performed to compare the performance of our slurry with commercially available slurries for stock removal and final polishing. Effects of additives in the slurry and effects of the choice of the seed on the polishing performance are also studied.

EXPERIMENT

Detailed procedures for the growth of silica particles and characteristics of grown silica particles were reported previously [4,5]. The primary particle size and morphology of silica were measured by a transmission electron microscopy (TEM, EM912, Car Zeiss). Microstructure of silica slurry was investigated using rheological measurement by ARES rheometer (Rheometric Scientific).

All the polishing experiments were done using a single platen, single-head chemical mechanical polisher (Model PM5) of Logitech Inc. A wafer is rotated about its center where being pressed down to a rotating pad by a carrier where colloidal silica slurry is injected on the pad. For our experiment, (100)-type, 4-inch wafers, of which one side is lapped, are used. Before polishing silicon wafer is cleaned with SC1 (Standard Cleaning 1, $NH_4OH : H_2O_2 : H_2O =$ 1:1:10) cleaning solution to remove particles on the wafer surface, rinsed with DI water and dried by nitrogen gas following previously published procedure [6].

Experimental conditions for the polishing are listed in table I [7]. In our study polyurethane-based Suba 600 made by Rodel Inc. is used as a pad. Each time after wafer polishing, pad is conditioned to reduce the glazing effect of pad which causes slower slurry delivery to the wafer surface and unstable and lower removal rate. Pad is conditioned by forming microscratches on the pad surface, thereby opens the pores of the pads and helps the flow of the slurry between the wafer surface and the pad [8]. Removal rate is calculated by the following equation,

$$\text{Removal rate} \left[\frac{\mu m}{\min} \right] = 10000 \frac{w_L}{\rho A t_p} \qquad (1)$$

Table I. Experimental condition for polishing

Specification	Stock removal polishing	Final polishing
Wafer carrier type	Flat	Flat
Platen rotation speed	70 RPM	40 RPM
Down load	3.17 psi	1.34 psi
Polishing time	30 min.	30 min.
Polishing pad	SUBA 600	UR 100
Polishing pad temp.	25 ~ 32 °C	25 ~ 32 °C
Pad conditioning	3 min. conditioning time & 10 min. relaxation time	

where w_L, ρ, A and t_p are weight loss(g), silicon density(g/cm^3), area of silicon wafer(cm^2) and polishing time(min).

RESULT AND DISCUSSION

Polishing slurry for stock removal usually has average particle size of 50 ~ 150nm and is recycled for over 500 minutes. Its main purpose is to smooth a wafer surface effectively by removing micro peaks on the silicon wafer. In the stock removal process, removal rate is more important than surface roughness, where surface roughness of 20 ~ 40Å is sufficient. Stock removal polishing slurry was prepared by adding polishing accelerator and dispersing agent to the silica slurry with particles grown from fumed or colloidal silica as seeds. Stock removal slurries made in this way were labeled as SRAG and SRLG according to the sort of seeds. SRAG and SRLG are slurries containing silica particles grown from Aerosil OX-50® and Ludox AS-40® as seeds, respectively. The average particle diameter of SRAG was much larger than that of SRLG as shown in Figure 1. Sizes of particles in the commercially available Slurry-N were found to be between those of SRAG and SRLG. Since silica particles are grown in a alcohol base solvent by Stöber method, alcohol in a slurry should be substituted with water for the use in a polishing slurry. Alcohol is evaporated by vacuum evaporation while distilled water is added to the slurry simultaneously. For more hardness of particles, prepared slurry was treated hydrothermally in the autoclave for 90 min at the temperature of 120°C.

Detailed compositions of additives and polishing rates are shown in table II. Removal rates with SRAG and SRLG are comparable to that of commercially available Slurry-N which has a removal rate around 0.35μm/min in our polishing test. Surface roughness of 20 ~ 40Å is sufficient in stock removal step. Surface roughnesses of the wafers polished with SRAG and SRLG are 29.7 Å and 19.1Å, respectively, which are better than the roughness value of 33.9 Å obtained with commercial Slurry-N in our test.

AFM image of lapped wafer before polishing is shown in Figure 2 (a), of which surface roughness is 1642 Å (scan area: 100μm × 100μm). AFM images of polished wafer using commercial Slurry-N, SRAG and SRLG are shown in Figure 2 (b)-(d), respectively (scan area: 5μm × 5μm). Even though the size difference between SRAG and SRLG is not small, noticeable differences are not seen in removal rate and surface roughness. This can be explained in two

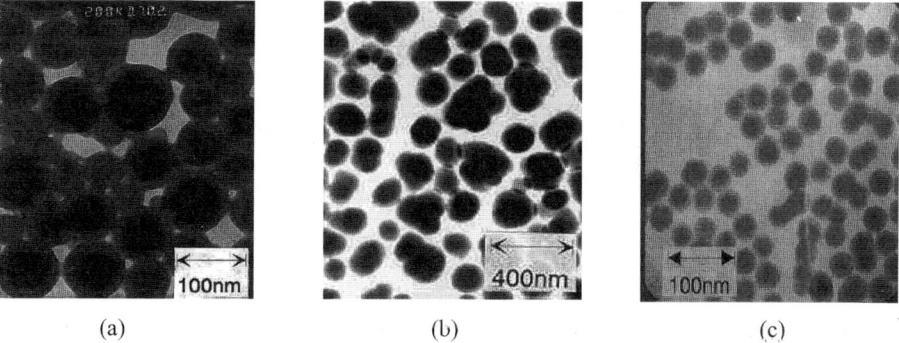

(a)   (b)   (c)

Figure 1. Shape of particles in the slurry; (a) Slurry-N(commercial), (b) SRAG and (c) SRLG.

Table II. Properties and polishing results of slurries

Additives	SRAG	SRLG
SiO$_2$(wt%)	11.5wt%, 1000ml	12wt%, 1000ml
Ethanol amine (ml)	35	35
TMAH (ml)	10	15
NH$_4$OH (ml)	0	10
Glycerol (ml)	5	5
Dilution pH	11.24	11.04
Removal rate (μm/min)	0.3097	0.3305

ways. First, since wafer polishing is carried out via chemical and mechanical actions between silica slurry and wafer surfaces, chemical contributions caused by additives such as polishing accelerator may be more dominant in polishing step than mechanical contribution which depends more on the abrasive particle size. Second explanation with respect to the stability of silica slurry may be more appropriate. That is, in spite of large differences in the primary particle size, silica particles of SRLG might form a flocculated structure and show similar polishing performances in removal rate and surface roughness. In order to clarify this possibility, rhelogical measurement was made using ARES rheometer. As shown in Figure 3, SRAG slurry shows stable Newtonian flow behavior up to 10wt% silica content indicating that SRAG slurry is composed of primary particles rather than flocculated structure. As silica content is increased to 20wt%, weak shear thinning behavior is observed. On the other hand, SRLG slurry (10wt% of silica) shows drastic shear thinning behavior and the slope of shear viscosity versus shear rate in log-log scale is −1 at low shear rate. This is indicative of the existence of yield stress and the formation of flocculated structure. The ready formation of flocculated structure of SRLG can explain the little differences in polishing performance between SRAG and SRLG. This simple comparison shows the importance of the dispersion stability of slurry particles in wafer polishing process.

Compared to stock removal polishing slurry, final polishing slurry has usually smaller

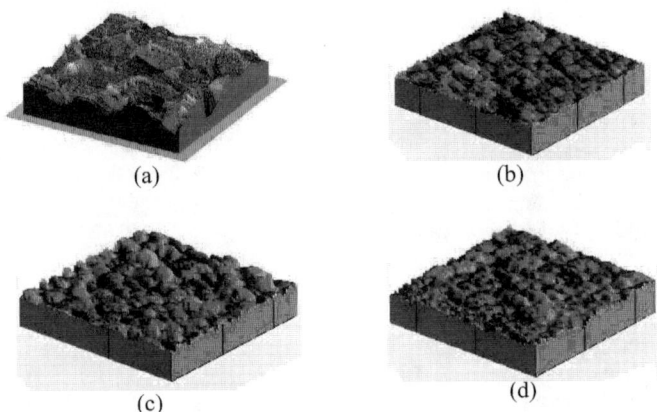

(a)   (b)

(c)   (d)

Figure 2. *AFM images of (a) lapped wafer and stock removal polished wafer using (b) Slurry-N, (c) SRAG and (d) SRLG, respectively.*

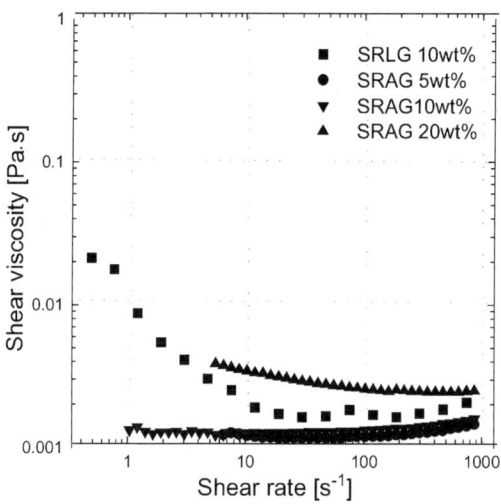

Figure 3*. Shear viscosity as a function of shear rate for various slurries.*

particle size and lower mass concentration for the sake of surface smoothness by sacrificing the high removal rate. Aerosil OX-50® can be used for final polishing slurry due to its high dispersion stability. However, since it has broad size distribution, particles of proper size should be separated to be used for final polishing slurry. Separated particles from Aerosil OX-50® by sedimentation are dispersed in D.I. water and stabilizer and accelerator such as cellulose, TMAH and ethanol amine are added. Cellulose makes a slurry flow to be laminar and TMAH and ethanol amine are used as pH controller, polishing accelerator and bactericide.

Effect of ethanol amine on the surface roughness of polished wafer was investigated by varying its weight fraction at 0, 0.011 and 0.033 wt % and the resulting RMS roughness were found to be 50.2, 7.67 and 33.8, respectively, in accordance with its concentration. Table III shows that roughness is getting better with small addition of ethanol amine but becomes worse with excess amount. It is because small amount of ethanol amine accelerates polishing but excess ethanol amine inhibits chemical reaction between slurry and wafer.

Effect of xanthan and cellulose shows similar tendency to that of ethanol amine. Table III shows that roughness becomes better with small addition of xanthan and cellulose but worse with more addition. Added amounts of xanthan and cellulose are same at 0, 0.005 and 0.01 wt %. From our results, minimum roughness can be expected with the concentration of additives at the vicinity of 0.01 wt % of ethanol amine or 0.005/0.005 wt % of xanthan/cellulose.

Table III. Effects of wt. % of ethanol amine, xanthan and cellulose on RMS roughness

Ethanol Amine (wt %)	Removal rate (Å)	Xanthan / Cellulose (wt %)	Removal rate (Å)
0.0	50.2	0.0/0.0	49.1
0.011	7.67	0.005/0.005	15.3
0.033	33.8	0.01/0.01	37.1

CONCLUSIONS

Performance of wafer polishing silica slurries and effects of additives on the performance were examined in this study. Silica particles for the stock removal polishing slurry was grown to proper size by sol-gel method using commercial fumed silica or colloidal silica as seeds and the resulting slurry was stabilized using electrostatic hindrance effect of surfactants. Colloidal silica (Ludox AS-40®) as seeds were grown to be 60 ~ 70nm in final size and particles showed more regular and mono-dispersed spherical shape than particles grown from fumed silica (Aerosil OX-50®). But, particles grown from fumed silica showed better stability after growth. This was confirmed by rheological measurement and gives the explanation of little difference in performance between the slurries containing large and small primary particles. Final polishing slurry was prepared by dispersing fumed silica in an alkaline solution and stabilized using surfactants' electrostatic and steric hindrance effect. Fabricated stock removal and final polishing slurries showed performance comparable to commercial ones. Study of the effects of additives showed the existence of optimal addition of additives for the best performance of the polishing slurry.

ACKNOWLEDGEMENTS

The authors wish to acknowledge support by KOSEF under Grant No. 1999-1-307-004-3 . This work was also partially supported by the Brain Korea 21 Project.

REFERENCES

1. W. Stöber and A. Fink, *J. Col. & Intf. Sci.*, **26**, 62 (1968).
2. G. H. Bogush, M. A. Tracy and C. F. Zukoski, *J. Non-Crystalline Solids*, **104**, 95 (1988).
3. S. Coenen and C. G. Kruif, *J. Col. & Intf. Sci.*, **124**, 104 (1988).
4. J.-H. So, M-H. Oh, J.-D. Lee and S.-M. Yang, *J. Chem. Eng. Japan*, *Submitted* (1999)
5. S. H. Bae, J.-H. So, S.-M. Yang and D. H. Kim, *J. Chem. Eng. Japan*, *Submitted* (1999)
6. F. A. Malik, U.S. Patent No. 5, 078, 801 (1992).
7. J. S. Basi and E. Mandel, U.S. Patent No. 4, 549, 374 (1985).
8. I. Ali and S. R. Roy, *Solid State Technol.*, **40**, (6) 185 (1997).

Mat. Res. Soc. Symp. Vol. 613 © 2000 Materials Research Society

Optical Characterization of Porous Membranes

Claudia Mujat, Lorrene Denney and Aristide Dogariu
School of Optics/CREOL, University of Central Florida
Orlando, FL 32816, U.S.A

ABSTRACT

Multiple light scattering techniques are intensively investigated as potential characterization tools for a broad range of applications. We are reporting on the noninvasive characterization of filters used in processes such as slurries filtering for CMP.

Filters are soft porous membranes characterized by their pore size distribution and thickness, and a noncontact, nondestructive optical procedure to measure these properties is highly desirable. Due to their internal inhomogeneity, porous media strongly scatter light and, therefore, a specific procedure needs to be developed.

In this work, low coherence interferometry is used to investigate light propagation in the filter and obtain the reflectivity as a function of optical pathlength for backscattered photons. This can be subsequently related to the optical properties of the sample using analytical and/or numerical models, and the porosity of the sample can be determined. In the case of filters with thicknesses much larger than the wavelength, a diffusion approximation for light propagation is used to infer the porosity information. For thinner membranes, numerical methods are used to describe the intermediate low-scattering regime that can not be represented analytically. As a direct result of the measurement, the thickness of the filter is determined independent of porosity.

INTRODUCTION

Porous media have a very important role in a wide variety of fields such as hydrology, chemical engineering, medicine and biology engineering. Many theoretical models and experimental methods have been developed to characterize their very complicated and irregular structure [1,2].

Here, we are investigating the possibility of using a multiple light scattering technique to determine the structural characteristics of the porous media. Light scattering studies were made on porous materials before [3], but the multiple scattering component was regarded as "noise" in these studies. However, the recent advances in the understanding of these phenomena, proved that multiple scattering is an important tool to investigate the microstructure of random media [4].

Some of the important parameters that characterize the structure of the porous materials, are the porosity P (defined as the fraction of the bulk volume of the porous sample that is occupied by pore space), the average pore size p_s, and the pore size distribution PSD. Since the pore systems consist of interconnected networks, one has to use a model in order to define and describe the pore size. The model is usually related to the method that is used to characterize the medium. For a multiple light scattering technique, the most intuitive and simple model is to consider the voids (pores) as spherical scattering particles. In this paper, we assess the possibility and the limits of using such a simplified model for porous media characterization.

EXPERIMENTAL RESULTS

The experimental technique is based on the principle of low coherence interferometry, and it uses a backscattering geometry to infer the optical pathlength distribution of the light reflected from the medium under investigation. This technique, called optical pathlength spectroscopy (OPS), was recently developed in our laboratories [5], and used for extensive structural investigations in random media such as colloids, powders, monitoring of slurries aggregation and sedimentation processes.

The experimental set-up is basically an all fiber optic Michelson interferometer. The pathlength resolved reflectivity from the sample is obtained by tuning the optical length of the reference arm and detecting the interference term. The light source is a broad bandwidth laser diode with a coherence length of 10μm, and a center wavelength of 1300nm. The same optical fiber is used to send light on the sample and to detect the light scattered by it. A more detailed description of the experimental set-up can be found elsewhere [5, 6].

The membranes under investigation are made by Millipore, and cover a wide range of materials (mixed cellulose esters, polyvinylidene fluoride and polycarbonate), thicknesses (10 - 150μm) and pore sizes (0.4 – 1.2μm). Each sample was placed under the measuring arm of the apparatus and the pathlength resolved bakscattering signal was recorded. To decrease the noise given by specular reflections, the reflectance signal was averaged over multiple successive scans.

First, the thickness was measured by placing the membrane on an X-Y translation stage and taking measurements of the reflection from the membrane and from the glass. Since the reflection is pathlength resolved, the distance between the two reflection peaks gives the thickness of the membrane (see figure 1).

The experimental values obtained for the average thickness are very close to the ones given by the manufacturer (table I). The only limitation is the coherence length of the source, because this determines the precision at which two points can still be separated in the optical path domain. This limitation is better reflected for the very thin sample (10μm) where the error is large.

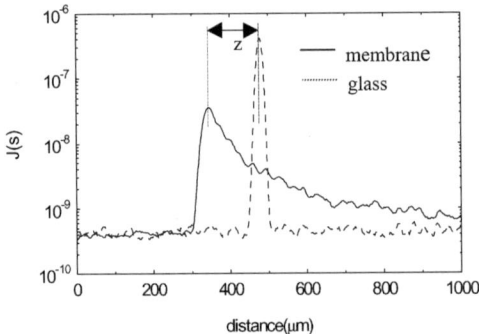

***Figure 1**. The distance between the two peaks gives the membrane thickness.*

Next, the OPS signal is recorded for the membrane dry and wet (figures 2 and 3). The purpose is to find an optical parameter that characterizes the medium that can be related to the pore size and implicitly to the porosity.

Table I. Measured vs. mean thickness indicated by the manufacturer

Membrane type	Mean thickness (μm)	Measured thickness (μm)
RA (mixed cellulose esters)	150	138
HVLP (polyvinylidene)	125	127
HTTP (polycarbonate)	10	20

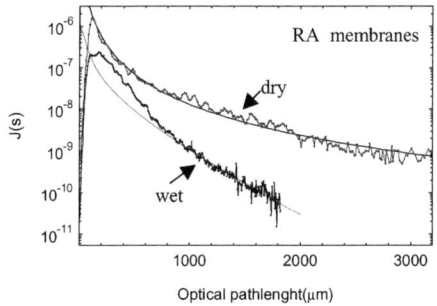

Figure 2. *Reflectance measurements of the RA type membranes in two measurement conditions dry and wet, as indicated. For the dry case, the curve can be fitted with equation (2), while for the wet case we used equation 3 to fit the data.*

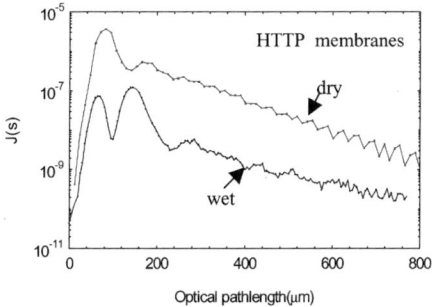

Figure 3. *Reflectance measurements of the HTTP type membranes in the two measurement conditions dry and wet, as indicated; numerical experiments can be used to recover these dependencies.*

RESULTS INTERPRETATION – DIFFUSION APPROXIMATION AND NUMERICAL SIMULATION

When the membrane is dry, the air bubbles inside the membrane pores are strongly scattering, while for the wet case, the refractive index contrast is decreasing and we have less scattering. As it can be seen in figure 2, the OPS signals for the wet and dry case are very different, and the interpretation of the results must be carefully considered.

One of the widely used analytical methods to describe the light propagation inside a random media is the radiative transfer equation. Since an exact solution of this equation is very difficult to obtain, the diffusion approximation is often used to get the optical properties of the medium under investigation. This approximation is valid only when the sample dimensions are much larger than the scattering mean free path, which is defined as the average distance between two scattering events. The diffusion approximation has been used to model different measurement geometries, and analytical expressions for the pathlength resolved reflection and transmission of the sample were calculated [7,8].

The diffusion process is described by the transport mean free path l_t that is essentially the path that the light travels inside the medium until its direction is randomized. This characteristic of the medium is related to the number density ρ of the scattering particles, the scattering cross section σ_s for a single scattering event, and the anisotropy g of the scattering particle, by the following relation:

$$ 1 = \frac{1}{\rho\sigma_s(1-g)} \tag{1} $$

Both the scattering cross section and the anisotropy depend on the particle size and the index contrast inside the medium. So, if the transport mean free path is measured for two different optical contrasts inside the same medium, the particle size can be determined. Mie scattering theory can be used to calculate σ_s and g for different particle sizes, under the assumption that the particles are spherical, and the scattering events are uncorrelated.

In the experimental geometry and in the limits of the diffusion approximation, the expression for the pathlength resolved reflectivity has the form:

$$ J(s) = A\, l_t^{-3/2} z_e\, s^{-5/2} \exp\left(-\frac{3z_e^2 l_t}{4s}\right) \tag{2} $$

where A is a constant that depends on the source strength, s is the distance that the light traveled inside the medium, and z_e is the extrapolation coefficient introduced to account for reflections at the boundaries. This equation can be applied to fit the experimental pathlength resolved reflectivity if the absorption is negligible and the ration between the thickness of the sample d and the transport mean free path l_t is much larger than 1.

For all the other cases when the thickness becomes comparable with l_t, the validity of this approximation is questionable. Further improvement can be done by considering the medium as a diffusive slab of finite thickness. The pathlength-resolved reflection from a slab [7, 8] can be expressed like:

$$ J(s) = B \sum_{m=-\infty}^{\infty} [(2md + 4mz_e l_t + l_t)\exp(-\frac{3}{4}\frac{(2md + 4mz_e l_t + l_t)^2}{l_t s}) - $$
$$ (2md + (4m-2)z_e l_t - l_t)\exp(-\frac{3}{4}\frac{(2md + (4m-2)z_e l_t - l_t)^2}{l_t s})] \tag{3} $$

where B is a constant depending on the source strength. The number of terms that have to be taken in the sum depends on the optical characteristics of the sample and on the fraction d/l_t. When this method fails too, a numerical experiment needs to be performed to recover the pathlength-resolved reflectivity for the given thickness.

The experimental data that we obtained is very different, not only from membrane to membrane (see figure 2, 3: RA vs. HTTP membrane), but also for different membrane conditions (dry vs. wet). The thickness of the membrane acts as a cutoff, and becomes an important parameter for the membrane characterization.

For the RA membrane, which is the thickest (130μm), the OPS signal recovered in the dry case is the one typical for a semi-infinite medium, can be fitted with equation 2 and an lt = 10μm is obtained. RA wet, however, has the trend given by equation 3. Due to the index contrast decrease inside the wet membrane, we have less scattering in the membrane and thus, a higher transport mean free path. A better fit is expected using equation 3 for the pathlength-resolved reflectivity. For very thin membranes, however, this equation is of limited use. Starting from values of l_t larger that d/3 a value for the transport mean free path can not be reliably obtained.

For the HTTP membrane, which is the thinnest (10μm), even in the dry case one can not make an analytical description and a numerical approach needs to be considered.

To recover the experimental geometry, photons are launched one by one into the medium in a pencil beam configuration, normally incident on the slab. First, an absorption length is randomly selected from an exponentially decaying distribution. The photon trajectory is then simulated, both scattering and reflections at the boundaries being taken into account. The process starts from the origin of the coordinate system and the first scattering event takes place along the oz axis. The distance between two scatterings is sampled from an exponentially decaying distribution characterized by the mean scattering length. The azimuthal angle ϕ is chosen to be any angle between 0 and 2π while the polar angle θ is sampled from the Henyey Greenstein distribution because it has a simple analytical form that can be used to express both isotropic and forward peak distributions by modifying the anisotropy factor. Reflection at the boundary is included as a function of the incidence angle θ by means of Fresnel equations averaged over both polarization states. We keep track of the photon pathlength and if it becomes larger than the predetermined absorption length, we discard it. For this particular experiment, the absorption length was considered negligible. All the optical properties of the medium are variable. The preliminary simulation results for d =130μm at different l_t's (figure 4) follow the pattern of the experimental results. As we can see, when the transport mean free path increases and

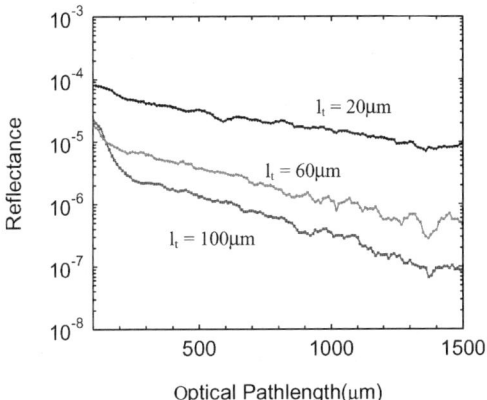

Figure 4. *Numerical results for a slab of 130 m thickness and different values of the transport mean free path.*

becomes comparable to the thickness of the slab, the pathlength resolved reflectivity curve becomes steeper. Since the change in the transport mean free path can be related either to different index contrast inside the medium (i.e dry vs. wet), or to different porosity (i.e RA vs. HTTP membranes), this kind of simulations can be used to match the experiment and in this way determine the optical properties of the sample, and implicitly the average pore size and porosity. As the pathlength dependence is determined by the ratio between slab thickness and pathlength, the results in figure 4 can easily be scaled to different slab thicknesses.

CONCLUSIONS

In this work, low coherence interferometry is used to investigate light propagation in a porous material and to obtain the pathlength-resolved reflectivity of the sample. This distribution can be subsequently related to the optical characteristics of the sample that change for different average pore sizes and implicitly, porosities.

The samples under investigation are porous membranes, whose finite thickness becomes another important parameter in their characterization, since in some cases this acts as a cutoff of the path that the photons can travel inside the membrane.

In the case of membranes with thicknesses much larger than the transport mean free path, a diffusion approximation for light propagation can be used to infer the porosity. For thinner membranes, only numerical methods can be used to describe the intermediate low-scattering regime that can not be represented analytically. As a direct result of the measurement, the thickness and porosity are determined independently.

BIBLIOGRAPHY

1. F.A. L. Dullien, "Porous Media - Fluid Transport and Pore Structure", Academic Press, San Diego, 1992
2. J. Rouquerol, F. Rodriguez-Reinoso, K.S.W. Sing and K.K. Unger (ed.), "Characterization of Porous Solids III", Elsevier, The Netherlands, 1994
3. Luca Cipelletti, Marina Carpineti and Marzio Gioglio, "Microporous membrane filters: a static light scattering study", Physica A, **235**, 248, (1997)
4. A. Ishimaru, "Wave Propagation and Scattering in Random Media", Academic Press, New York, 1978
5. G. Popescu and A. Dogariu, "Optical pathlength spectroscopy of wave propagation in random media", Optics Letters, **24**, 442, (1999)
6. G. Popescu, C. Mujat and A. Dogariu, "Evidence of scattering anisotropy effects on boundary conditions of the diffusion equation", Phys. Rev. E, **61**, (2000)
7. M.S. Patterson, B. Chance and B.C. Wilson, "Time resolved reflectance and transmittance for the non invasive measurement of optical tissue properties", Applied Optics, **28**, 2331, (1989)
8. D. Contini, F. Martelli, and G. Zaccanti, "Photon migration through a turbid slab described by a model based on diffusion approximation. I. Theory", Applied Optics, **36**, 4567, (1997)

CMP Consumables

Mat. Res. Soc. Symp. Vol. 613 © 2000 Materials Research Society

Interfacial Fluid Pressure and Its Effects on SiO$_2$ Chemical Mechanical Polishing

C.Zhou*, L. Shan*, J. R. Hight*, S.H. Ng*, A. J. Paszkowski**, J. Tichy*** and S. Danyluk*

*School of Mechanical Engineering, Georgia Institute of Technology, Atlanta, GA30332
** Chemical Products Corporation, Cartersville, GA
*** Mechanical and Aerospace Engineering Department, Rensselear Polytechnic Institute, Troy, NY

ABSTRACT

In this paper, the experimental results of interfacial fluid pressure and friction measurements during polishing are presented, as well as their dependence on some major process variables. Under simulated conditions, a sub-ambient fluid pressure was observed, and its magnitude was of the same order of magnitude as the applied polishing load. Since this fluid pressure is non-uniformly distributed, the contact stress, obtained by combining the effects of both applied load and the fluid pressure, is not uniform across the wafer and will result in non-uniform material removal. The mechanism of the presence of the fluid pressure was investigated, and an analytical model was developed to predict the magnitude and distribution of this fluid pressure. The effects of the sub-ambient fluid pressure on material removal rate and profile were tested with thermally grown silicon dioxide on 100mm diameter, P-type (100), single crystal silicon wafers. The polishing experiments show the effect of sub-ambient fluid pressure on polishing rate and profile.

INTRODUCTION

With increasing integration density, especially the emergence of new materials and technologies, chemical mechanical polishing (CMP) has become an indispensable step in the fabrication of integrated-circuit chips. CMP is being used in process architectures to form shallow trench isolation, metallize gates, form Cu-based interconnect and low "k" dielectrics [1], and so on. Current IC technology requires a global planarity of 0.3μm, which is still achievable with the most advanced polishing equipment [2]. However, the increase of IC operating frequencies requires better site flatness and microroughness so that the signal distortion and clock skewness caused by dielectric thickness variation may be minimized [3]. CMP needs to be improved to meet even more stringent requirements.

Mechanical interactions, such as contact stress and fluid pressure, are of extreme importance in wafer polishing, especially for the planarity and microroughness of the finished surfaces. But there is little literature on the fluid (slurry) pressure at the polishing interface, which is a very important factor[4]. First of all, from the lubrication stand point, knowledge of the interfacial fluid pressure may yield the nature of the contact. The fluid pressure may also affect the slurry transportation at the interface and contribute to the contact stress that dominates the polishing process. To investigate the fluid pressure at the interface, an experimental apparatus was constructed to measure the fluid pressure.

The measurements were used to obtain the large area and magnitude of sub-ambient pressure, and the overall effect of a suction force in addition to the applied normal load. These forces are expected to affect the polishing rate and uniformity [4].

This paper addresses the study of mechanical interactions during CMP and develops a theory that can be used as a guide for CMP process design. The effects of process variables on this interfacial fluid pressure were studied, including fluid viscosity, pad structure and modulus, velocity, normal load, pad surface roughness, and wafer surface curvature. In addition, shear forces at the interface were also measured. The friction coefficient was about 0.1-0.4, which indicates contact at the pad/silicon interface. The effects of the sub-ambient fluid pressure on material removal rate and polishing uniformity were tested with silicon wafers.

EXPERIMENTS

A. Experiments for sub-ambient pressure measurement

The apparatus shown in Figure 1 was constructed using a commercial bench top polishing machine (Logitech model PM4) with custom-built additions. The polishing machine has a 305mm (12inch) diameter stainless steel turntable driven by a 1/3 horsepower DC motor. Speed can be controlled to within 1 rpm from 5 to 70 rpm and the turntable can withstand 450 N vertical load as well as 450 N horizontal (frictional drag) load. Polishing pads were bonded onto the turntable with double-sided pressure sensitive adhesive. In most cases, a commercially available, impermeable, void filled, cast polyurethane polishing pad (Rodel IC1000) of approximately 1.28mm thickness was used. A custom-built overhead structure was added to support the samples and to mount the instrumentation. Load was applied to a fixture through a linear bearing shaft, and a ball joint was used to allow the fixture to follow the pad. The load shaft housing was supported horizontally by two load cells used to measure the friction (shear) force at the interface. Both load cells have a capacity of 2000N with a resolution of 0.2 N. During the experiment, water or a glycerin solution was applied instead of the slurry. The pressure of the fluid trapped at the 'wafer/pad' interface was measured with the data acquisition system. One of the pressure sampling fixtures is a 100 mm diameter, 25 mm thick, stainless steel solid disk with 16 – 0.4 mm diameter holes drilled through to the 'pad/wafer' interface. The array of holes can be aligned and fixed at any desired direction during the polishing simulation. The surface of the pressure fixture was lapped to ~10μm for a center/edge difference making the surface convex. The fluid pressure was measured by connecting the holes to pressure sensors with a range of −103 to +103kPa and a resolution of 0.1 kPa, and the signal was recorded with a computer controlled data acquisition system. To test the effects of surface curvature/profile on the fluid pressure, another pressure sampling fixture was designed with a 2.5mm thick stainless steel plate supported by a 6mm steel ring as shown on the left of Figure 2. The surface curvature/profile of the plate can be adjusted with a set screw and the normal load is applied onto the bridge structure over the steel ring. The center deflection (CD) of the plate was recorded for different surface curvatures/profiles, which was measured using a capacitance probe with an accuracy better than 0.1μm. The polishing pad surface has an average roughness of about 5μm after being conditioned with an 80μm diamond-grit tool. For comparison, a sheet of mylar tape was attached on the top of the pad, which gave an average surface roughness of about 0.5μm and reduced the friction coefficient from 0.4 to 0.1. During most pressure measurements, the pad was flooded with water, and the fixture was held stationary.

A. Experiments on wafer polishing

One hundred millimeter diameter, p-type (100) oriented silicon wafers were thermally oxidized to produce an oxide thickness of 1.3 μ m. These wafers were then polished on a Struers RotoPol-35 tabletop polishing machine with a RotoForce-3 head. The rotating speed of the RotoForce-3 head was 150rpm. A schematic diagram of the polishing setup is shown in Figure 3. The polisher was modified to include a peristaltic pump to deliver 50ml/min slurry to a consistent location on the pad. The wafer holder consisted of a retaining ring, carrier film and an oxidized silicon wafer. The carrier film (Rodel Part #A06132) is elastic with a self-adhesive back. The holder disc was machined to a flatness of 25 μ m, then polished by standard metallographic techniques using alumina slurries until the flatness was approximately 5 μ m center-to-edge, as measured with a Hommelwerke profilometer.

Rodel IC1000 pads and ILD 1300 slurry were applied for the experiments. Since the sub-ambient fluid pressure tends to increase with speed, the material removal rate was measured under constant load (20kPa) and time (2 minutes) at various relative speeds, including 0.4m/s, 0.6m/s, 0.9m/s, 1.4m/s, 2.0m/s, and 3.0m/s. The pads were conditioned with a 160 μ m grit diamond-coated disk, which produced an average pad surface roughness of 6 μ m. The pads were conditioned in this manner before the start of each experiment. The reduction in the oxide thickness after polishing was determined by a Plas-Mos Ellipsometer. The average silicon dioxide thickness of three points near the center of the wafer was used to calculate the polishing rate.

RESULTS AND ANALYSIS

The typical magnitude and distribution of the interfacial fluid pressure is shown in Figure 4, i.e. a large portion of the interface exhibits a negative (sub-ambient) pressure, and a small positive pressure exits around the trailing edge. The average fluid pressure at the interface is "negative" and acts as a suction force on the order of 50~100% of the applied normal load.

The effects of process variables, such as normal load, relative velocity, pad surface roughness and modulus, fluid viscosity, and target surface curvature, were studied by comparing the 1-D fluid pressure distribution (tangential with respect to polishing platen). Figure 5 shows the effects of relative velocity and pad surface roughness on the interfacial fluid pressure. With the increase of the relative velocity, the magnitude of the sub-ambient pressure increases, and the zero point shifts backward. Also, the magnitude of the sub-ambient pressure increases with the increase of pad surface roughness, and the effect is significant. Figure 6 shows the effects of pad modulus and fluid viscosity at a velocity of 0.16m/sec and 20kPa normal load. For higher pad modulus (24MPa vs 12Mpa measured for this compression test), the magnitude of the sub-ambient pressure is slightly smaller, and almost no positive pressure can be observed at the tail. Fluid viscosity also has an effect on the interfacial fluid pressure, i.e. the higher the viscosity, the larger the magnitude of the sub-ambient pressure. Figure 7 shows the effects of target surface curvature on the interfacial fluid pressure and the pressure sampling fixture, for the IC1000 perforated pad, 20kPa normal load, and 0.6m/sec relative velocity. The effect of the surface curvature on the fluid pressure is significant, i.e. with the surface changed from flat to a 100μm CD convex, the fluid pressure at the leading half of the interface changed from sub-ambient to positive, and that at the trailing half flattened out to zero.

Based on the above observations, an analytical model was developed to physically explain the experimental results. Figure 8 shows the contact and fluid mechanics at a polishing interface. First, the contact stress between a rigid flat (wafer) and a compliant pad is not constant but has a bowl shape, i.e. high stress at the edge and low stress at the center [5]. With the motion and a shear force at the interface, the contact stress distribution is skewed. Since the pad is much rougher than the wafer, the applied normal load is actually supported by the pad asperities, and there exists a liquid film at the interface [6], which varies with the contact stress, i.e. a thicker film at the center and a thinner film at the edge. From lubrication theory, a negative fluid pressure is generated at a divergent gap and a positive fluid pressure is generated at a convergent gap [7]. This is the reason why a negative pressure was measured at the leading edge and positive pressure at the trailing edge of the sample. Using the above physical model, the interfacial fluid pressure along the tangential central line can be estimated by use of the models and equations presented in reference [4].

Figure 9 shows an example of the magnitude and distribution of the calculated fluid pressure vs. location on the sample. The modeling results fit the measured values reasonably well. Refer to reference [4] for more details of the modeling work.

Since the fluid pressure was measured with a stationary fixture, while most commercial CMP's have rotational wafer holders, a friction test was performed to determine the difference between the two conditions. Figure 10 shows the result of the test. The rotation of the fixture tends to increase the interfacial friction, which indicates a larger suction force (higher average sub-ambient pressure). Therefore, holding the fixture stationary will not result in exaggerated sub-ambient pressure effect. The results of polishing an oxide film are shown in Figure 11. The rotating speed of the fixture with wafer was 150rpm. The material removal rate (MRR) versus the radial distance on the wafer shows that the MRR is reasonably constant (within 700 and 800 Å/min.) over most of a 100mm diameter wafer except in the last five millimeters. Near the edge, the polishing rate increase increases by ~30%. Other data of MRR versus relative speed are also shown in Figure 11. The MRR is non-linear with speed, increasing at a greater rate than linear.

DISCUSSION

According to the experimental results and the above analysis, there exists a non-uniform interfacial fluid pressure that generally gives a sub-ambient pressure at the leading edge and a positive pressure at the trailing edge. The existence of this sub-ambient fluid pressure tends to "pull" the two surfaces (pad/silicon) into intimate contact and may have two major effects on the material removal during CMP. First, it may affect the slurry transportation at the polishing interface. The sub-ambient pressure will improve the slurry flow into the interface, while the positive pressure tends to "squeeze out" the slurry from the trailing edge. This pressure distribution may also result in a "driving force" that resists the backward flowing of the slurry and retards the slurry replenishment. Secondly, as mentioned before, the magnitude of the fluid pressure is of the same order of magnitude as the applied normal load. Therefore, the addition of this fluid pressure may significantly change the distribution of the contact stress. Figure 12 shows a free-body force analysis of the fixture. The resulting contact stress $\sigma(x)$ is non-uniform and tends to result in non-uniform material removal. This pressure distribution will influence material removal in the following way: using Preston's equation as an example [9], the material removal rate (MRR) is proportional to applied unit load and the relative velocity:

$$MRR = K \cdot P \cdot V$$

For a non-uniform contact stress, P is the unit load caused by the localized contact stress, i.e. $P = \sigma(x)dx$. The overall contact stress $\sigma(x)$ may be expressed as:

$$\sigma(x)dx = P_{Applied} + p(x)dx$$

where p(x) is the fluid pressure and $P_{applied}$ is the localized contact stress caused by the applied normal load. Therefore,

$$MRR = K \cdot [P_{Applied} + p(x)dx] \cdot V$$

and the contribution of the fluid pressure to the material removal of stationary wafer polishing is shown in the above equation. However, for most CMP processes, because of the rotation of the wafer, the polished surface may not follow the exact shape of the contact stress $\sigma(x)$ distribution as shown in Figure 12, but the trace of this non-uniformity should be seen.

Figure 11(a) shows the polishing profile indicating the non-uniformity of the polished surface. The first three points close to the center at different polishing speeds were averaged and plotted in Figure 11 (b). According to Preston's equation, the material removal rate should be proportional to the polishing speed, which is shown as the dashed line in Figure 11(b). However, the polishing results deviate from the Preston prediction with speed, and this indicates that the sub-ambient fluid pressure is acting as an additional load and increasing the polishing rate, or,

$$MRR = k*(P_{applied} + P_{fluid}) * V$$

where P_{fluid} increase with speed V.

CONCLUSIONS

1) The leading two-thirds of the wafer/pad interface exhibits a "negative" fluid pressure, and the trailing third a positive pressure during CMP process. The magnitude of this pressure can be on the order of the applied load, and thus it could result in a non-uniform polished surface.

2) Both magnitude and distribution of this interfacial fluid pressure varies with several process variables, such as fluid viscosity, pad structure and modulus, velocity, pad surface roughness, and wafer surface curvature.

3) Because of the sub-ambient fluid pressure, the MRR is not proportional to applied unit load and the relative velocity (Prestonian Equation). The sub-ambient fluid pressure may act as an additional load and increase the polishing rate.

REFERENCES

[1] D. Pramanik, M. Weling and X.W. Lin, 1998, *"CMP Applications for Sub-0.25μm Process Technologies"*, Proceedings of the Second International Symposium on Chemical Mechanical Planarization in Integrated Circuit Device Manufacturing, v98-7, pp 1-8.

[2] The International Technology Roadmap for Semiconductors: 1998 update

[3] B. E. Stine, R. Vallishayee, *"On the impact of dishing in metal CMP processes on circuit performance"*, Int Workshop Stat Metrol Proc IWSM 1998 IEEE Piscataway NJ USA, p64-67

[4] L. Shan, J.Levert, and S. Danyluk, 1999, *"Interfacial Pressure Measurements in Chemical Mechanical Polishing Interface"*, MRS Symposium Pro., v566, 187-195

[5] K. L. Johnson, *"Contact Mechanics"*, Cambridge University Press, Cambridge, UK, 1985, p41

[6] J. A. Greenwood and J. B. P. Williamson, 1966, *"Contact of Nominally Flat Rough Surfaces"*, Proc Roy Soc London, A295, p300-319.

[7] B. J. Hamrock, 1994, *"Fundamentals of Fluid Film Lubrication"*, McGraw-Hill, p141.

[8] N. Patir and H. Cheng, 1978, *"An Average Flow Model for Determining Effects of Three Dimensional Roughness on Partial Hydrodynamic Lubrication"*, ASME Journal of Lubrication Technology, Vol. 100, No. 1, p12-17.

[9] F. W. Preston, *"The Theory and Design of Plate Glass Polishing Machines"*, Journal of the Society of Glass Technology, v11, 1927, p214

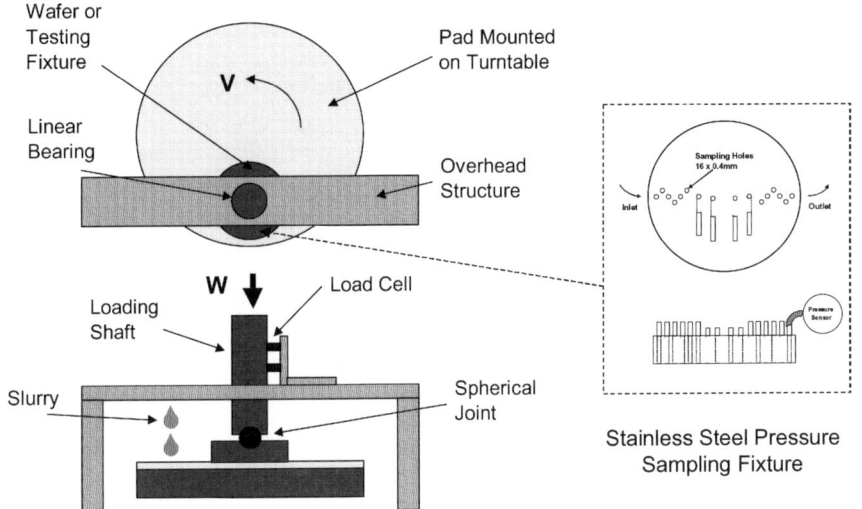

Figure 1. Schematic of the polishing platen and overhead structure

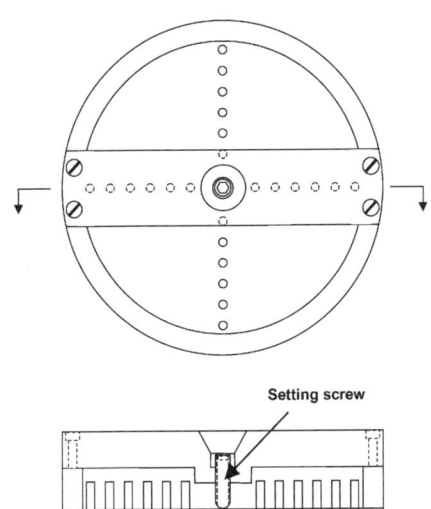

Figure 2. Pressure sampling fixture with adjustable surface curvature

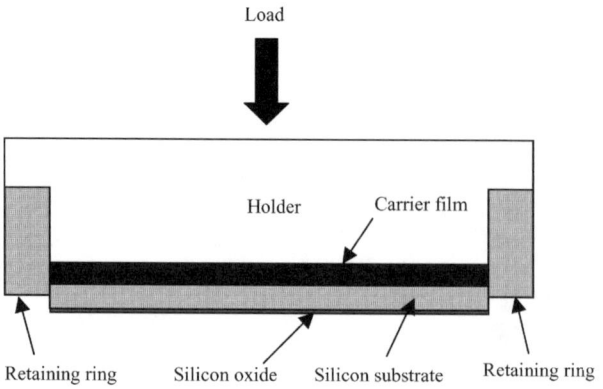

Figure 3. Cross-sectional view of the wafer holder

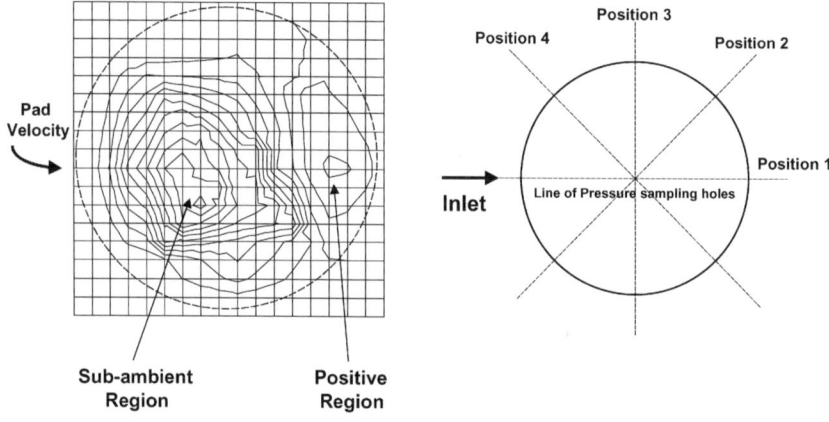

Figure 4. 2-D isobaric mapping of interface fluid pressure

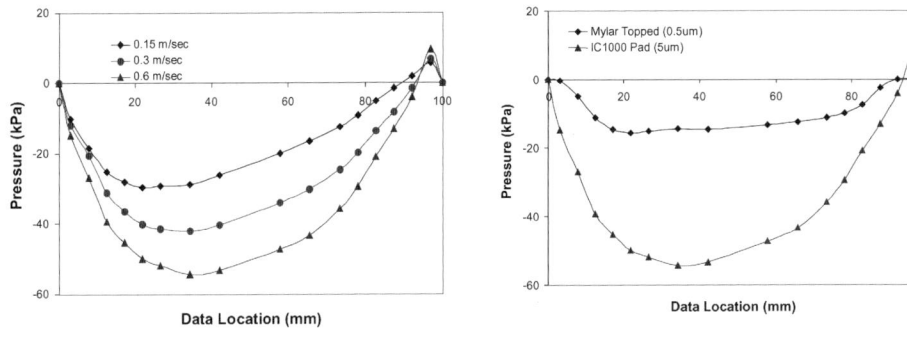

(a) Effect of Relative Velocity

(b) Effect of Pad Roughness

Figure 5. The effects of relative velocity and pad surface roughness.
IC1000 perforated pad, 20kPa normal load

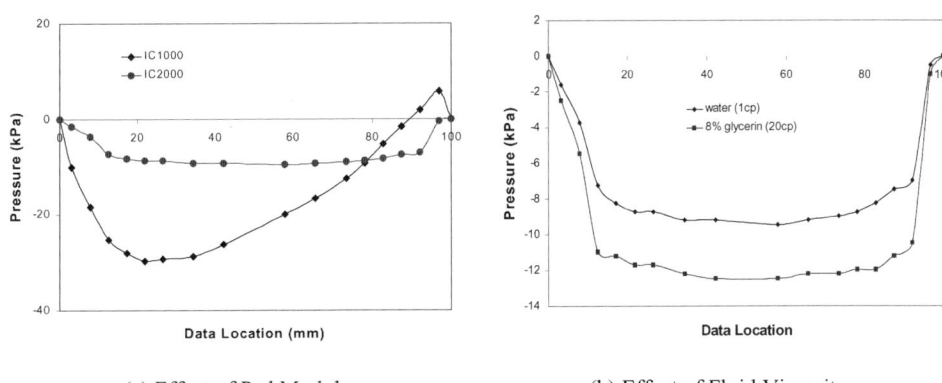

(a) Effect of Pad Modulus

(b) Effect of Fluid Viscosity

Figure 6. The effects of pad modulus and fluid viscosity
20kPa normal load and a velocity of 0.16m/sec

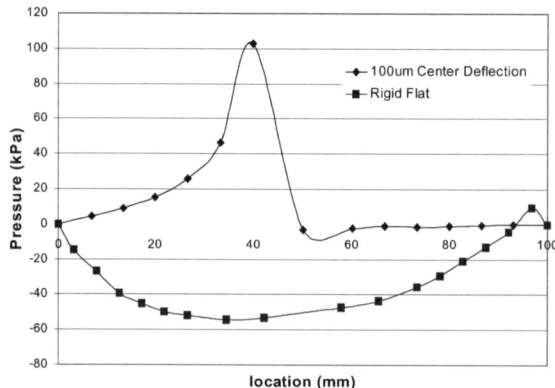

Figure 7. The effects of surface curvature on the interfacial fluid pressure.
IC1000 perforated pad, 20kPa normal load, and 0.6m/sec relative velocity

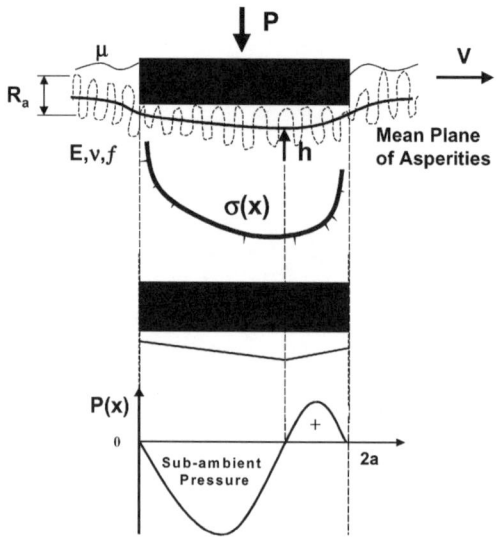

Figure 8. Schematic of the physical system (with compressed
asperities), contact stress, and predicted pressure distribution.

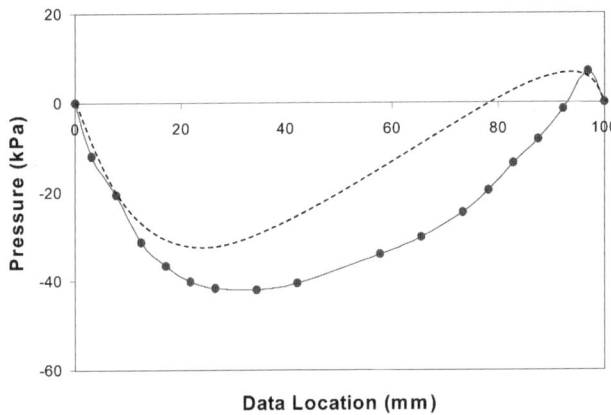

Figure 9. The calculated fluid pressure vs. measured curve.
IC1000 perforated pad, 20kPa normal load, 0.3m/sec velocity

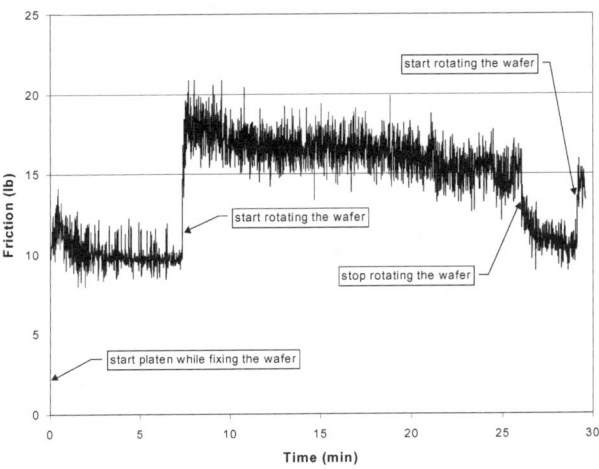

Figure 10. Friction test on the effect of fixture rotation
IC1000 pad, 0.3m/sec velocity, 20kPa normal load

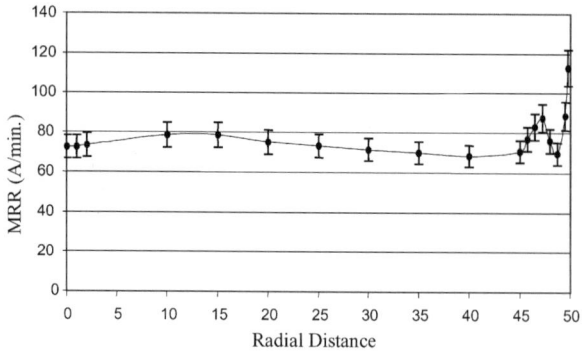

(a) Polish profile of wafer at 20kPa and 0.6m/s

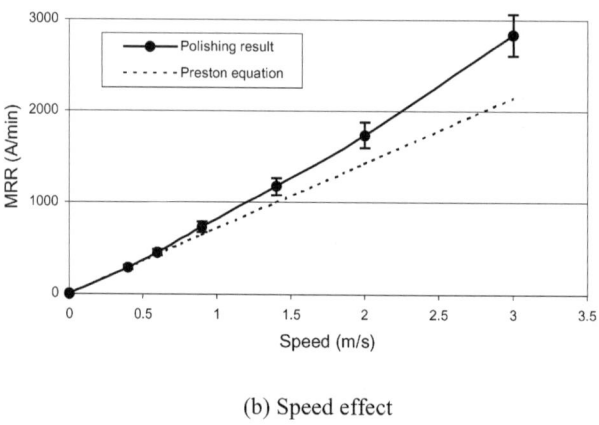

(b) Speed effect

Figure 11. The material removal rate versus (a) radial distance and (b) relative speed.

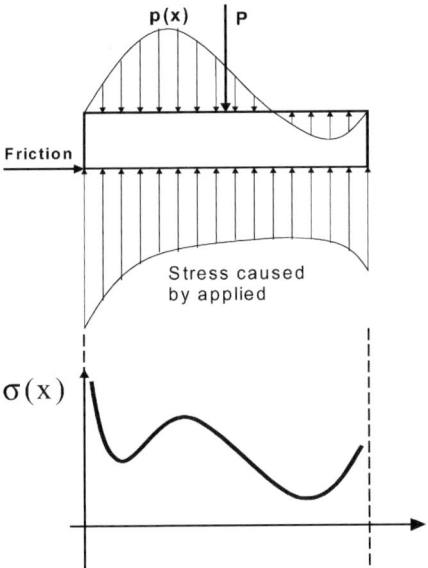

Figure 12. The addition of interfacial fluid pressure to the applied normal load results in a non-uniform contact stress

Mat. Res. Soc. Symp. Vol. 613 © 2000 Materials Research Society

Dynamic Mechanical Analysis (DMA) of CMP pad materials

Irene Li[1], Kersten M. Forsthoefel*, Kathleen A. Richardson[1], Yaw S. Obeng[2], William G. Easter[2] and Alvaro Maury[2]
[1] School of Optics/CREOL, Department of Chemistry, University of Central Florida, 4000 Central Florida Blvd., Orlando, FL 32816, USA.
[2] Bell Laboratories, Lucent Technologies, 9333 South John Young Parkway, Orlando, FL 32819, USA.
*now at Univ. of Pennsylvania, Chemistry Department

ABSTRACT

In the semiconductor industry, there is a need to establish fundamental, mechanism-based, correlation(s) between process conditions, consumables (e.g., pads and slurries), and observed process performance in Chemical-Mechanical Polishing (CMP). In this paper, we present recent results of studies on polyurethane-based CMP pads in static and dynamic conditions using Dynamic Mechanical Analysis (DMA) to monitor modulus and energy damping changes. Two-layered, composite IC1000 on Suba IV pads, [IC1000 (cast and cured polyurethane elastomer) / Suba IV (polyurethane impregnated polyester felt)], were analyzed: prior to use (fresh); after a 24-hr soak in silica-containing oxide slurry (basic); and after oxide polishing (used). Upon comparison it was observed that a characteristic transition feature due to water is present at sub-ambient temperatures in both the slurry soaked and used pads. Exposure of as-received pads to basic oxide slurry results in a broad, high temperature transition thought to be the result of chemical-induced disruption of macrostructural units. Polishing (load-enhanced chemical exposure) introduces further changes to the polymer network represented by an apparent degradation to the structural species responsible for the high temperature transition in Suba IV.

INTRODUCTION

Based on previous experiments examining pH and solvent absorption, and its impact on Shore hardness [1], the following model was proposed to interpret the interaction behavior observed between the polyurethane pads and its chemical environment. Upon exposure, the solvent or buffer wets, penetrates, and swells the polyurethane matrix. Once inside the matrix, solvent or any other dissolved nucleophilic species (e.g., OH⁻) can attack at the carbonyl center (-NC(O)O-) of the urethane structure, thus destabilizing the polymer matrix. Such a structural change should be detectable using mechanical analysis techniques.

As CMP pads undergo both chemical and mechanical interaction during the actual polishing process, this study compares the changes of the pad matrix upon exposure to a simulated chemical/mechanical environment. Both static and dynamic pad exposure experiments were carried out using Dynamic Mechanical Analysis (DMA) to probe material changes in as-received (new) polyurethane pads. Common basic oxide polishing slurry was used as the soak medium for static experiments. A used pad measured following oxide polishing in the same slurry, was analyzed to compare the effects of mechanical load as well as chemical exposure.

To probe potential changes in pad microstructure that might result from the suggested nucleophilic attack on the urethane functional group, we used Dynamic Mechanical Analysis (DMA) to monitor modulus (stiffness) and damping (energy dissipation) of pad samples, pre- and post- slurry exposure.

The fundamental operation of DMA is based on Equation (1):

$$G^* = G' + i\,G'' \qquad (1)$$

where G^* is the complex modulus, G' is the storage modulus, G'' is the loss modulus, and i is the imaginary root of -1. Storage modulus (G') is measured directly and represents the elastic properties of the material and examines a material's ability to return or store energy. The loss modulus factor (iG'') is related to the viscous properties of the material and its "liquid-like" qualities. In other words, this term describes energy dissipation by the material. The ratio of loss modulus to storage modulus (tan delta [δ] or damping) is a ratio of energy dissipated per cycle to the maximum potential energy stored during a cycle [2,3]. By using these different moduli, DMA allows improved understanding of a material's thermal and mechanical properties.

EXPERIMENTAL PROCEDURE

Changes in static and dynamic pad properties of as-received (new), slurry soaked (static), and pad samples that have seen polishing (dynamic) were examined by DMA using a TA Instruments DMA 2980, (TA Instruments, New Castle, DE). Sample strips with a geometry of about 15 mm long x 5.5 mm wide x 1.25 mm thick, were run in tension at 1 Hz at an amplitude of 10 μm in Autotension mode at 120%. The holes of the IC1000 side were aligned in the middle of the mounted part and specimens were consistently cut with the same number of holes positioned identically. Samples were mounted in the DMA at room temperature and then cooled to $-125°C$ (torqued to ~6-8 psi), fitted with a heat shield, and held isothermally for approximately 10 min prior to ramping. Samples were then run from $-125°C$ to $200°C$, at a ramp rate of $5°C$/min.
In an attempt to differentiate the influence of chemical interaction with the combined effect of chemical exposure under mechanical load, static testing on IC1000/Suba IV was done by soaking the pads, IC1000 side down, in a beaker of slurry for 24 hrs. At the allotted time, the sample was removed with tweezers, rinsed with DI water to remove excess slurry, and blotted dry in preparation for DMA. Dynamic specimens (used in polishing experiments) were removed from the polishing platen and held in deionized water prior to analysis. The samples were then prepared as stated above.

RESULTS & DISCUSSION

In discussing the DMA data, we focused primarily on the energy dissipation (tan δ) of the pad samples. Transition temperatures (error within $\pm 6°C$) were determined from the peak of the tan δ curve [4]. These damping curves are characterized by peak heights and shapes that change systematically with increasing amorphous phase, and are related phenomelogically to glass transition temperature (T_g) [3]. For example, sharp transitions represent changes in well-defined "crystalline" domains within the pad matrix. On the other hand, broad, ill-defined traces represent changes in "amorphous" or plasticized domains within the pad matrix.
Multiple runs were done to verify reproducibility of the data, and to determine the minimum number of scans (2-3) necessary for obtaining representative spectra. Figure 1 is an overlay of DMA scans on individual IC1000 and Suba IV pads. IC1000 (green-solid curve), as expected, is a stiffer material than the Suba IV (blue-dashed curve) backing material, as can be seen by the higher storage modulus. Although there is an overlap in tan delta transition temperature, as

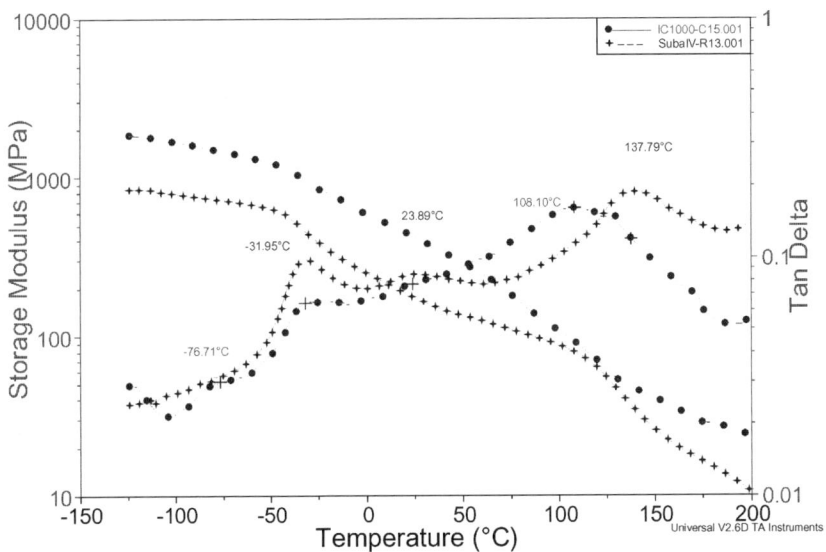

FIGURE 1. *DMA overlay of individual IC1000 (green-solid curve) and Suba IV (blue-dashed curve) scans. IC1000 has a higher storage modulus (stiffness) and a tan δ transition at T ~ 105°C, and ~ −80°C. In Suba IV, its distinguishing transition is at a higher temperature (T ~ 140°C), but it also exhibits a transition at ~ 25°C. The two pads share a similar transition at T ~ −30°C.*

would be expected since both pads contain polyurethane in their make-up, each pad has its own inherent characteristics (transition T and peak shape) unique to its material morphology (e.g., foam versus impregnated felt). IC1000 and Suba IV have transitions at sub-ambient temperatures of −80° and −30°C, and −30°C, respectively. Suba IV has a broad transition at 25°C and a sharper feature at 140°C, whereas IC1000 shows a transition only at 105°C.

Figure 2 is a scan on an as-received (new) IC1000/Suba IV pad sample. Since the material is a composite of IC1000 and Suba IV, the features in the stacked pad reflect the presence of both materials. The ill-defined features at lower temperatures are a resultant effect of IC1000 and the Suba IV sub-pad. The three "soft" shoulders seen below zero (T ~ -75, -30, -2, and 20°C) are most likely associated with small, local (i.e., side chain) structural rearrangement. As the free volume (the space a molecule has for internal movement) increases upon heating, whole side chains and localized groups of four to eight backbone atoms begin to have enough space to move. Thus, the 110°C transition is most likely due to larger scale, macrostructural (i.e., polymer backbone) modification that has absorbed sufficient energy to initiate such movement. The broadness of this transition's range encompasses transitions seen in individual IC1000 (T ~ 105°C) and Suba IV (T ~ 140°C), respectively. However, it is apparent that the IC1000 transition dominates in this region. Typical transition position variation for repeated runs of "identical" samples is within 6°C.

In Figure 3, a DMA scan of a stacked pad following 24 hr slurry soak illustrates the chemical effect of introducing the pad to a basic environment. The "soft shoulders" previously seen in the as-received pad sample at low temperatures have disappeared and an intense, distinguishing transition appears at ~ 10°C. The freezing and subsequent thawing of water from the slurry, results in this characteristic modification at sub-ambient temperatures during the DMA run.

The extremely broad transition that we have labeled at ~125°C suggests the presence of a large number of plasticized or "amorphous" domains within the pad matrix, possessing a broad distribution of similar, structural configurations. We have no way of distinguishing in the composite pad structure what aspect of the transition comes from IC1000 versus Suba IV, but most likely both pad materials contribute. The breadth of the transition is indicative of a range of various "types" of molecular configurations, requiring a range of activation energies for structural rearrangement. The generation of such a broad distribution of sub-units during slurry exposure, supports our proposed aqueous attack mechanism which leads to chemically induced changes in the pad matrix; destabilization (breakdown) of the pad matrix results from

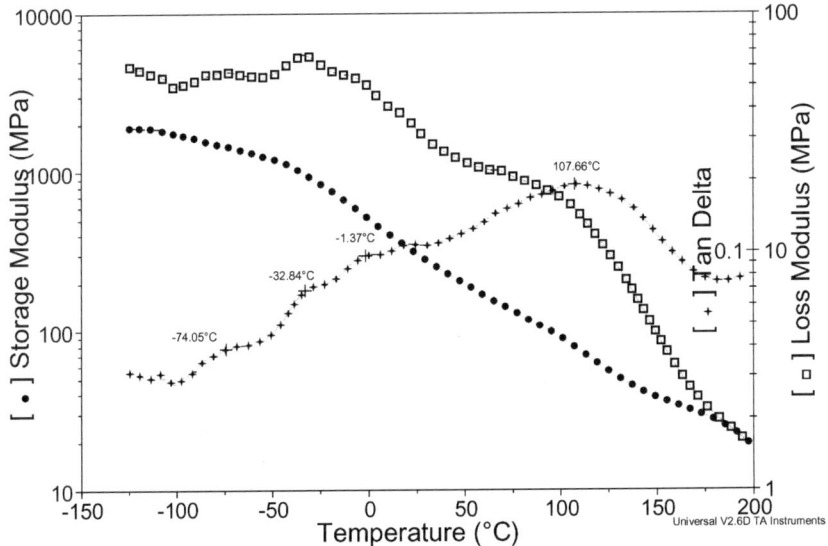

FIGURE 2. *DMA scan of an as-received (new) IC1000/Suba IV stacked pad. The sub-ambient temperature transitions in tan δ are due to small, localized bond movements. The broad peak at ~ 110°C results from main chain movements and is representative of primary transitions (i.e., T_g).*

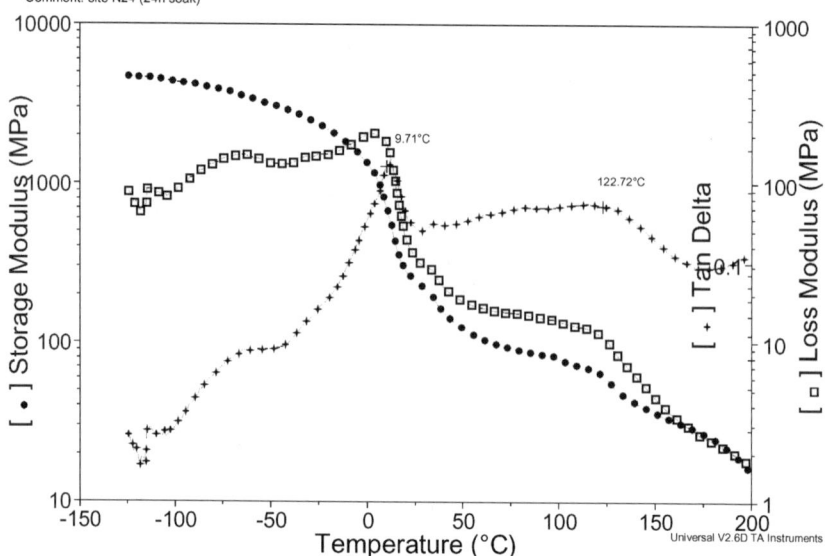

FIGURE 3. *DMA scan of a stacked IC1000/SubaIV pad following a 24 hr static slurry soak. The basic aqueous environment has modified the sub-ambient tan δ transitions seen in the as-received sample, resulting in a sharp peak at ~10°C. The wide transition range (~125°C) in the soaked sample is broader than in the as-received pad.*

nucleophilic attack by slurry OH⁻ species at the carbonyl center (-NC(O)O-) of the urethane structure.

Figure 4 is a scan of an IC1000/Suba IV pad after oxide polishing. The peak at 113°C is much sharper than the broad peaks previously seen at similar temperatures in Figures 2-3. The polishing process (chemical exposure *with* mechanical wear) produced a more pronounced feature at 113°C, suggesting a narrower distribution of structural units, and a reduction in contribution of units associated with the T ~140°C transition for Suba IV, seen in Figure 1. We propose that the resulting feature (T ~ 113°C) is solely attributable to the IC1000 layer suggesting that the Suba IV matrix and hence, its DMA signature, has been significantly modified by the chemical / mechanical polishing environment. Experiments to investigate the mechanical contribution to the degradation excluding chemical interaction are ongoing.

DMA

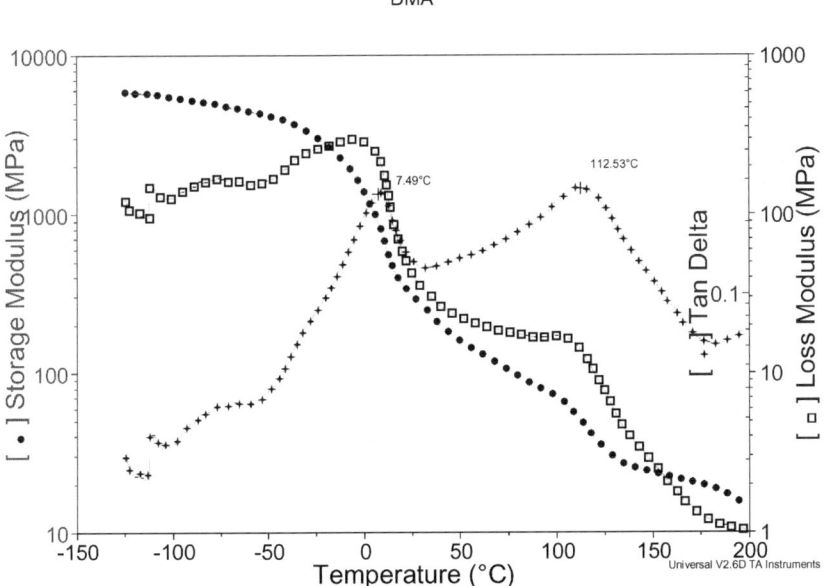

Figure 4. *DMA scan of a stacked pad after undergoing oxide polish. Two dominant transitions appear at 8°C and 113°C. The former is due to the uptake of water from the slurry during the polishing process; the latter is a direct consequence of chemo-mechanical induced structural modification to the pad matrix.*

Further evidence to support Suba IV degradation and the loss of the 140°C transition can be seen in considering the effect of slurry exposure to the polyurethane contained in the Suba IV felt structure. While slurry enters and leaves the composite pad structure through the porous IC1000 top layer, the fibrous Suba IV structure can "wick" the slurry, swelling the felt structure allowing retention of the aggressive suspension. Such retention provides extended exposure to the interwoven polyurethane polymer, allowing degradation to progress. Such sustained exposure, could result modifications of the sub-pad, leading to the disappearance of structural features responsible for the transition. This model is consistent with independent Shore hardness tests on individual pad components that show little change in hardness to IC1000, but a measurable softening in Suba IV, upon identical slurry exposure.

Figure 5 summarizes the tan δ transitions for the three cases illustrated previously, for as-received pads and those undergoing static chemical exposure and the chemical/mechanical impact of polishing. In addition to the transition changes highlighted earlier, a measurable difference in storage modulus, or stiffness, is clearly evident for the static and spent pad samples. The higher modulus in the upper traces (static soak and after use pad) is due to water freezing in the pad structure. The sudden drop in modulus once the water leaves the structure (in the 0-100°C range) reveals a clear loss of stiffness at higher temperatures, being most notable in the worn pad which saw polishing. This supports the theory that it is not only chemical exposure, but rather exposure under load that serves to degrade the pad with use. These data highlight the potential applicability of DMA to quantify such subtle microstructural differences that accompany the wear of pads in the CMP process.

CONCLUSIONS

The effects of static and dynamic slurry exposure on the mechanical properties of IC1000/Suba IV pads were investigated using dynamic mechanical analysis (DMA). Individual pad components contribute to a combined DMA signature for the composite pad, which undergoes changes when exposed to chemical and chemical/mechanical environments. Exposure of as-received pads to basic oxide slurry results in a broad, high temperature transition thought to be the result of chemical-induced disruption of macrostructural units. Polishing (load-enhanced chemical exposure) introduces further changes to the polymer network represented by an apparent degradation to the structural species responsible for the high temperature transition in Suba IV.

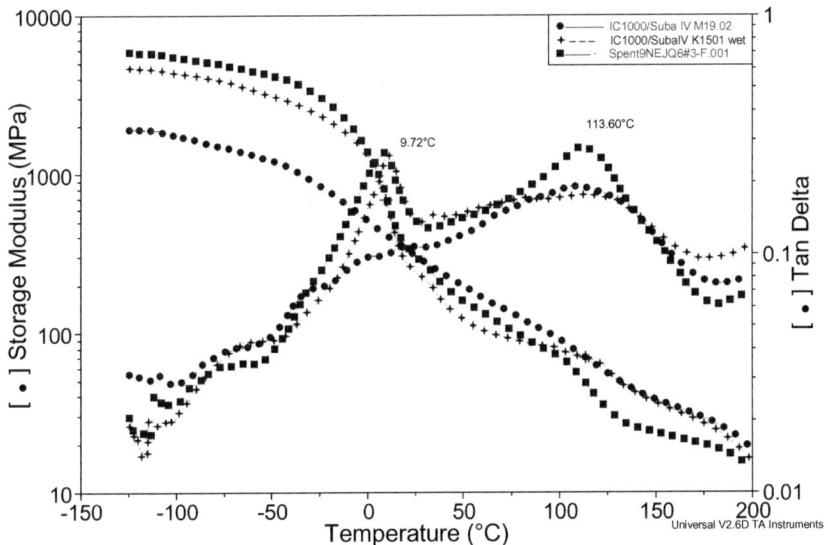

Figure 5. *An overlay of as-received (blue), slurry-soaked (green), and used (red) polishing pads (IC1000/Suba IV) highlighting the influence of chemical interaction, and when it is combined with mechanical load.*

These data uphold our proposed model that a solvent (i.e., slurry) facilitates nucleophilic attack, via solvation, of the polymer backbone. Chemically, the slurry disrupts the polyurethane network, but when used in actual polishing where mechanical force plays an integral role, the ability to purely chemically modify the structure is impeded. By further probing the individual contribution of the mechanical and chemical components of the polishing process, on the pad materials, we can improve our understanding of pad wear, and extend pad lifetime.

ACKNOWLEDGMENTS

The authors would like to thank the following people and institutions for their technical, financial, and material support: Lucent Technologies, Rodel Corporation, and Mr. Jeff Groh of TA Instruments.

REFERENCES

1. I. Li, K. M. Forsthoefel, K. A. Richardson, and Y. S. Obeng, Mechanistic aspects of the relationship between CMP consumables and polishing characteristics, *Thin Solid Films*, submitted for publication (Jan 2000).
2. K. P. Menard, *Dynamic Mechanical Analysis (A Practical Introduction)*, CRC Press, Boca Raton, FL (1999) pp.4.
3. E. Turi, *Thermal Characterization of Polymeric Materials*. 2nd Ed., Vol 1. Academic Press, Brooklyn, New York (1997) pp.486
4. R. B. Prime, J. M. Burns, M. L. Karmin, C.H. Moy, and H-B. Tu, *J. of Coatings Tech.*, **60**, No. 761, 55-60 (1988).
5. J. M. Steigerwald, S. P. Murarka, R. J. Gutmann, *Chemical Mechanical Planarization of Microelectronic Materials*, Wiley-Interscience Publication, New York (1997) pp. 66-67.

Mat. Res. Soc. Symp. Vol. 613 © 2000 Materials Research Society

An Evaluation of the Effects of Benzotriazole in NH₄OH Slurry for Copper CMP

V.S.C. Len, D.W. McNeill and H.S. Gamble
School of Electrical & Electronic Engineering, The Queen's University of Belfast, Belfast BT9 5AH, Northern Ireland

ABSTRACT

Chemical mechanical polishing (CMP) of copper using alumina-based NH₄OH slurry containing benzotriazole (BTA) has been evaluated in terms of polish efficiency and viability. Dishing of damascene copper patterns can result from a combination of chemical dissolution and mechanical abrasion due to the deformed polishing pad bending into the recessed copper regions. The addition of at least 0.1 wt.% BTA to the slurry leads to the formation of a thin Cu(I)-BTA polymer on the copper surface during CMP. This polymer reduces the amount of dishing by an order of magnitude. At the same time, however, the CMP polish rate falls sharply with the addition of 0.1 - 0.25 wt.% BTA to the slurry. Above 0.25 wt.% BTA, the polish rate falls no further. Stability of alumina particles in the NH₄OH slurry is found to deteriorate with the addition of BTA. Integrated copper/barrier electromigration resistance test structures with large contact areas (2x2mm) have been successfully patterned using a 2-step CMP/etching process scheme, using a BTA-containing slurry to minimise dishing.

INTRODUCTION

CMP has become an essential technique for patterning copper multi-level metallisation (MLM) as device sizes continue to shrink below the current 0.18-micron technology. Problems such as dishing in copper lines and erosion of the inter-layer dielectric are major obstacles in achieving a successful damascene copper process. The copper in the recess trenches is susceptible to dishing, especially during the overpolish step of at least 10%. Thus, some preventative measure is desirable to protect the copper surface in the trench areas during polish steps.

Pioneering copper CMP work using alkaline and acidic slurries with oxidants such as oxygen, ferricyanide ($(Fe(CN)_6^{3-})$ and hydrogen peroxide (H_2O_2) was carried out and reported by Steigerwald et al. [1,2,3]. Other copper CMP slurry chemistries in various media and problems related to these chemistries were investigated by Carpio et al. [4]. Luo and co-workers [5,6] studied the incorporation of BTA in the slurry and the related slurry stability in acidic and alkaline media.

BTA works as an inhibiting agent by reacting with the copper surface or native oxide to form a Cu(I)BTA polymer. This polymer can prohibit water, acids and alkalis from the metal surface [7] and offers excellent corrosion protection in harsh conditions [8].

The viability of copper CMP using alumina-based NH₄OH slurry with and without the addition of BTA is examined. A brief study of the stability of alumina particle suspension in both slurries is also presented.

EXPERIMENTAL DETAILS

A modified Lapmaster lapping machine with a 14" diameter Suba IV/IC1000 stacked pad was used to perform copper CMP. The platen was driven at a fixed rotation speed of 60 rpm and the superimposed wafer carrier rotation was also 60 rpm. Aqueous slurry was delivered to the rotating platen at a constant rate of 130 ml/min. A fixed applied pressure of 1.75 psi was used in all polish tests. The polishing slurry consisted of $2 - 4$ wt.% α-alumina abrasives (300 nm particle size) added to $0.1 - 0.75$ wt.% of BTA in 5 vol.% NH_4OH. The slurry was continuously agitated during the polishing process to ensure complete dispersion of the mixtures.

For damascene patterning tests, 1.5 μm layers of silicon dioxide (SiO_2) were grown by wet oxidation on 100 mm n-type silicon substrates. Trenches of different line widths were produced by RIE of the SiO_2 to a depth of 1.2 μm. The trenches were then overfilled with a $30 - 50$ nm layer of sputtered titanium followed by a $1.3 - 1.5$ μm layer of copper without breaking vacuum. Test structures consisted of a matrix of trench line widths and spaces. Line width/space densities of 10, 33, 50, 67 and 83% were generated using trench line and space widths between 2 and 200 μm. For copper electromigration resistance structures, line widths between 0.9 and 2 μm and contact pads of 2000x2000 μm were produced using the same method.

Copper polish rate was determined by measuring the change in the sheet resistance of the copper films at regular intervals during polishing, using a 4-point probe. Dishing and surface profiles were observed by an Alpha-step 2000 profilometer and a Nanoscope III atomic force microscope (AFM) operated in tapping mode.

EXPERIMENTAL RESULTS AND DISCUSSIONS

Polish Rate

To establish the effectiveness of BTA as a copper protector for damascene copper CMP, copper-coated patterned wafers were polished in NH_4OH slurry with various concentrations of BTA in the slurry. The end results were compared to those polished without BTA. Polish conditions were fixed using 5 vol.% NH_4OH slurry containing 2 wt.% alumina, a pressure of 1.75 psi with constant pad maintenance and BTA concentration varying from $0.1 - 0.75$ wt.%. The polishing rates of BTA-containing slurries are depicted in Figure 1. Copper polish rate plunges from 400 nm/min to 65 nm/min with the addition of 0.1 wt.% BTA. The change in polish rate as BTA concentration in the slurry is increased above 0.1 wt.% is less significant. At 0.25 wt.% BTA the polish rate has fallen to 42 nm/min, but further increase in BTA concentration has little effect. The results suggest that a 0.1 wt.% BTA addition to a NH_4OH slurry is adequate to form a protective layer of Cu-BTA polymer on the copper surface which retards mechanical abrasion and chemical dissolution.

On the higher regions, the polish pad and the abrasives in the slurry try to mechanically abrade away this monolayer as soon as it forms. For the 0.1 wt.% BTA it is likely that the removal of the Cu-BTA complexes allows some copper oxidation and chemical dissolution of the abraded copper to take place faster than the polymer formation. At BTA concentrations of 0.25 wt.% and above, the BTA reaction appears to be sufficient to stop chemical dissolution of the abraded copper by either not allowing copper oxide to form or by complexing the copper oxide before it can form soluble complexes with the ammonia. The polish process becomes almost entirely mechanical and, thus, no longer a sensitive function of BTA concentration. Polish rates are degraded but improved planarisation is achieved.

However at BTA concentrations of 0.5 wt.% and above, the abraded material is no longer solubly complexed by the ammonia. After 2 - 3 min, the asperity contact between the pad and wafer decreases as the pad surface becomes saturated with both Cu_2O and Cu-BTA complexes. This impedes the transport of slurry to the wafer and hence polish rate drops.

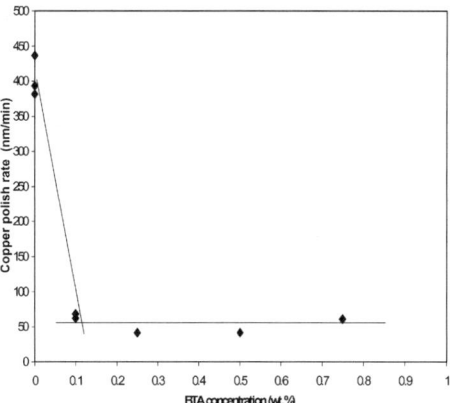

Figure 1 *Copper polish rates versus BTA concentration (5 vol.% NH_4OH, 2 wt.% alumina, 1.75 psi applied pressure).*

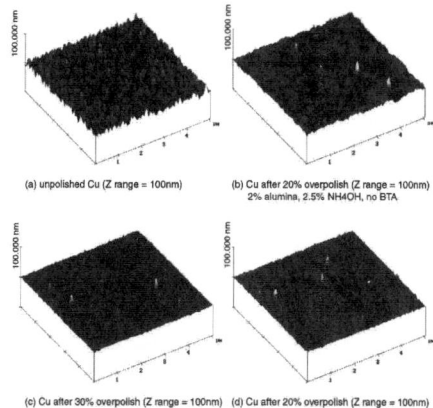

(a) unpolished Cu (Z range = 100nm)

(b) Cu after 20% overpolish (Z range = 100nm) 2% alumina, 2.5% NH4OH, no BTA

(c) Cu after 30% overpolish (Z range = 100nm) 2% alumina, 5% NH4OH, 0.5% BTA

(d) Cu after 20% overpolish (Z range = 100nm) 4% alumina, 5% NH4OH, 0.25% BTA

Figure 2 *AFM scans of surface roughness of Copper surfaces (a) as-deposited, (b) CMP with 2 wt.% alumina, no BTA, (c) CMP with 2 wt.% alumina with BTA, and (d) CMP with 4 wt.% alumina with BTA.*

To improve the polish rate, alumina concentration was increased from 2 to 4 wt.% using the same NH_4OH slurry. It was found that the polish rate improved slightly to 65 nm/min compared to the 42nm/min using 2% alumina. However, an increase in copper surface roughness was observed in terms of scratches and defect density.

The present experimental results show that the overall polish rate obtained with BTA present is much lower than that without BTA compared to other researchers [6]. The low polish rate may be attributed to the low platen speed and applied pressures together with the use of higher concentrations of BTA.

Surface Roughness

The AFM studies illustrated in Figure 2 (a) to (d) show the surface morphology of the copper surfaces before and after polish. The grainy as-deposited copper surface (Figure 2a) has been smoothed out after subsequent CMP processes (Figure 2b – d). The root-mean-square (RMS) roughness of unpolished copper was approximately 6.5 nm, and the value decreased to 3.3 nm for standard copper CMP in 2 wt.% alumina, 2.5 vol.% NH_4OH slurry without BTA. The RMS roughness was further decreased to 1.72±0.06 nm when the samples were polished with

BTA added to the slurry. However, in all cases, with and without BTA addition, where the Copper was polished, numerous scratches were evident. It is interesting to correlate these RMS roughness values with the corresponding visual images shown in Figure 2. Copper polished with NH$_4$OH slurry without BTA appears coarse and uneven as seen in Figure 2(b). This suggests that besides mechanical abrasion, the copper surface was also attacked by the slurry chemicals during the polish cycles via the grain boundaries. The latter images in Figure 2(c) and 2(d) exhibit a smoother and flatter surface which correlate well to their small RMS roughness values. It is postulated that the Cu-BTA surface film has protected the underlying copper from chemical dissolution effectively and the only surface irregularities are derived from scratches induced by the abrasives. The introduction of a higher alumina concentration of 4 wt.% caused more clusters of micro-scratches intermixed with random deep scratches as seen in Figure 2(d). Some of the surface scratches may have been caused by particles embedded onto the polish pad during the polish cycle as the polishing was carried out in a non-clean environment. Nevertheless, it may be concluded that an alumina abrasive concentration of 2 wt.% in the slurry is preferable to the higher concentration of 4 wt.%.

Dishing

Dishing of copper in the patterned recesses was measured after removal of copper from the surrounding field areas (defined as the endpoint), and also after an overpolish step ranging between 10 and 30% of the polish time to endpoint. At endpoint, the remaining thin titanium adhesion layer in the central region of the 4" wafer was removed by dipping briefly in dilute HF before dishing values were measured. After the titanium removal in the HF etch, the copper lines may protrude slightly on the wafer surface, yielding negative dishing values at endpoint under certain circumstances. After a 10% overpolish, negative dishing is no longer observed.

Figure 3 illustrates that a significant reduction in the amount of dishing in recessed areas has been achieved with the incorporation of BTA for all pattern densities. In the absence of BTA, the normal dishing values vary between 400 and 600 nm for all pattern densities after a 10% overpolish. With a BTA concentration of only 0.1 wt.% and 10% overpolish, the dishing values have dropped to less than 150 nm in all cases. Further increase of the BTA concentration to 0.25 wt.% reduces the dishing values to less than 30 nm, more than an order of magnitude less than the values obtained without BTA. There is some evidence that higher concentrations of BTA in the range 0.5 – 0.7 wt.% may actually cause a slight increase in dishing again. This may be due to the poor dissolution of abraded copper in the slurry making the polishing process less efficient. For a 20% overpolish, without BTA the dishing values range between 700 and 850 nm, with 0.1 wt.% BTA they are in the range 100 - 200 nm, and with 0.75 wt.% BTA they are all approximately 100 nm.

The measured values may clearly be explained in terms of the formation of a Cu-BTA polymer on the copper surface. At a BTA concentration of 0.1 wt.%, the reduction in dishing is thought to be due to the formation of a discontinuous polymer layer. Polymer formation occurs at a slower rate than polymer removal by abrasion. At BTA concentrations of 0.25 wt.% and above, the polymer layer appears to be more continuous and dishing is effectively minimised. A BTA concentration of approximately 0.25 wt.% is, therefore, the optimum choice.

In Figure 4, the effect of overpolishing on dishing is illustrated. Without BTA in the slurry, dishing increases linearly from 350 nm at endpoint to 1000 nm after a 30% overpolish. With

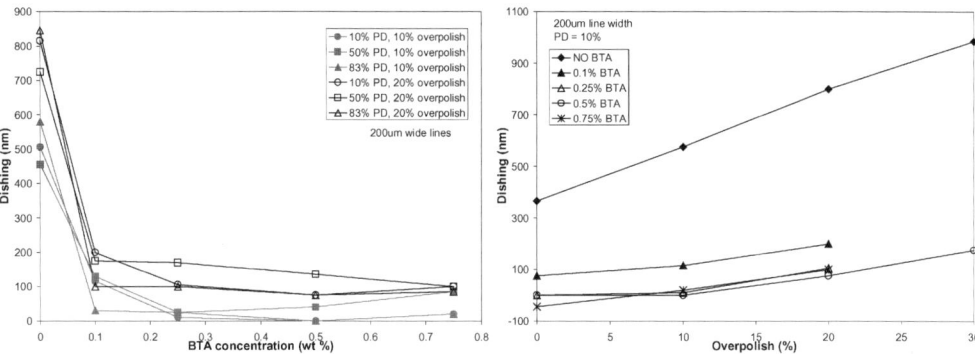

Figure 3 *Dishing behaviour on 200 μm wide lines with various concentrations of BTA, PD = 10 – 83%.*

Figure 4 *Effect of overpolishing on dishing using various BTA concentration in NH₄OH slurry. Line width = 200 μm, PD = 10%.*

BTA, not only is dishing less at endpoint, but it also increases less quickly during the overpolish. For example, with 0.5 wt.% BTA, dishing is still less than 200 nm even after a 30% overpolish.

CMP of copper patterns for electromigration tests, including large 2mm×2mm contact pads, was successfully developed using a 2-step process where copper CMP was done up to endpoint, followed by a brief 1% HF dip to remove the remaining titanium, and a final 10% overpolish. The optimal copper CMP recipe is 5 vol.% NH₄OH, 2 wt.% alumina (0.3μm) slurry added with 0.25 wt.% BTA, at 1.75 psi, using thin titanium (<30 nm) as an adhesion layer. Dishing was found to be reduced by an order of magnitude compared to polishing in slurry without BTA. BTA passivation of copper films may have advantages for subsequent processing, but if it is necessary to remove the Cu-BTA polymer, desorption is reported to be achievable at 199°C [8].

Stability of Alumina Abrasives in BTA Containing Slurry

To determine the settling rate of 2 wt.% alumina in 5 vol. % NH₄OH slurry with 0 – 0.75 wt.% BTA, a 125 ml clear PVC bottle was used to hold the slurry. 100ml of slurry was kept in the bottle and, prior to measurement, the bottle was shaken vigorously to deagglomerate the solid particles. The interface between the suspension and the clear liquid above was closely monitored over a period of time and used to determine the settling characteristics.

It was observed that for NH₄OH based slurry, with pH levels between 10.9 and 11.5, the suspension of 0.3μm α-alumina is very stable. Figure 5 shows the settling characteristics of alumina particles within the suspension over time. The vertical axis is a normalised indication of the height to which solid particles are still suspended, whilst the horizontal axis shows the time taken for the solid particles to settle to a particular height. The settling of alumina particles in a 5 vol.% NH₄OH slurry is minimal, at less than 3 mm over a period of 12 hr observations. With the addition of 0.1 wt.% BTA, however, the slurry becomes unstable and requires constant agitation to disperse the particles during polishing. The alumina particles settled rapidly during the first 3

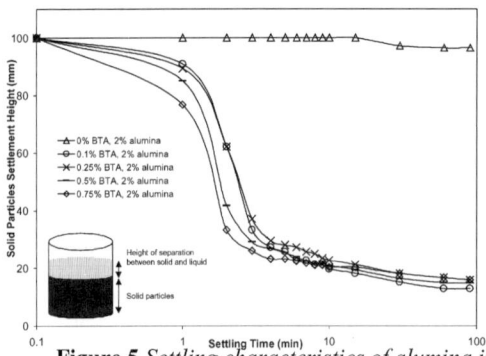

Figure 5 *Settling characteristics of alumina in 5 vol.% NH₄OH slurry over time.*

min and then gradually over the next 10 – 90 min. The stability of alumina slurry is derived from the electrostatic repulsion between the positively charged alumina particles. As BTA is added to the slurry, the negatively charged BTA⁻ ions become absorbed onto the alumina particles, neutralising their positive charge and reducing the electrostatic repulsion forces [5]. This allows the alumina particles to settle more quickly. The effect is very noticeable, even at a BTA concentration of 0.1 wt.%, and is exaggerated as the BTA concentration is increased.

CONCLUSIONS

Copper CMP has been achieved using NH₄OH-based slurry with and without the addition of BTA. In slurry without BTA, significant dishing is observed and copper thickness in the trenches is reduced considerably. However, with the addition of between 0.1 and 0.75 wt.% BTA, dishing is dramatically reduced. At the optimum value of 0.25 wt.%, dishing values were an order of magnitude lower than those obtained without BTA. Addition of BTA was accompanied by a reduction in polish rate, but this could be overcome in a commercial polisher. Copper films polished in a BTA-containing slurry had a much lower RMS surface roughness than those polished in normal slurry. Stability of alumina particles in NH₄OH-based slurries is poor when BTA has been added, and constant agitation of the suspension is required. The optimum BTA concentration in the current CMP work was found to be approximately 0.25 wt.% where there is a good balance between polymer formation and dissolution of abraded copper.

REFERENCES

1. J.M. Steigerwald, R. Zirpoli, S.P. Murarka, D. Price, R.J. Gutmann, *J. Electrochem. Soc.*, **141**, 2842 (1994).
2. J.M. Steigerwald, S.P. Murarka, J. Ho, R.J. Gutmann, D.J. Duquette, *J. Vac. Sci. Tech B*, **13**, 2215 (1995).
3. J.M. Steigerwald, S.P. Murarka, R.J. Gutmann, D.J. Duquette, *Materials Chem & Phys*, **41**, 217 (1995).
4. R. Carpio, J. Farkas, R. Jairath, *Thin Solid Films*, **266**, 238 (1995).
5. Q. Luo, D. R. Campbell, S. V. Babu, *Langmuir*, **12**, 3563 (1996).
6. Q. Luo, R. A. Mackay, S. V. Babu , *Chem. Material*, **9**, 2101 (1997).
7. R.R. Thomas, V.A. Brusic, B.M. Rush, *J. Electrochem. Soc.*, **139**, 678 (1992).
8. V. Brusic, M.A. Frisch, B.N. Eldridge, F.P. Novak, F.B. Kaufmann, B.M. Rush, G.S. Frankel, *J. Electrochem. Soc.*, **138**, 2253 (1991).

Mat. Res. Soc. Symp. Vol. 613 © 2000 Materials Research Society

Fundamental Study of Iodate and Iodine Based Slurries for Copper CMP

Seung-Mahn Lee, Uday Mahajan, Zhan Chen and Rajiv K. Singh
Department of Materials Science and Engineering and Engineering Research Center for Particle
Science and Technology, University of Florida, Gainesville, FL 32611

ABSTRACT

The chemical mechanical polishing of copper in several slurry chemistries based on iodate and iodine oxidizers has been investigated. Benzotriazole (BTA) and potassium iodide (KI) were used for preparing the polishing slurry chemistries based iodate. As observed by the anodic electrochemical behavior of copper and the surface analyses of EDS, it was determined that CuI layer formed in the iodate and iodine based solutions. Especially, in I_2 slurry in pH 4, CuI layer formed very fast and uniformly, and passivated the copper. In addition, the highest removal rate using this slurry was obtained. These results were compared to H_2O_2 based slurries. From these experimental results, the slurry containing 0.1M KIO_3 and 0.01M KI, and 0.01N I_2 give better results than H_2O_2 based slurry in copper CMP.

INTRODUCTION

Recently, copper has emerged as an attractive material for interconnect applications, as compared to aluminum, with close to half the resistivity (reduced RC time delay and the number of levels of metal required) and with improved reliability due to electromigration resistance on the order of 10 times higher [1-4]. Therefore, Copper CMP has now been recognized and accepted as the process that is capable of providing the planarity to build multilevel interconnects schemes with below quarter-micron lines. One of the key issues in copper CMP is the development of slurries which can provide high removal rates, good planarity and high selectivity [5]. A number of slurry chemistries have been studied to optimize copper CMP conditions: HNO_3 and H_2O_2 as oxidizers [3,6], benzotriazole (BTA) as an inhibitor [3,4], NH_4OH as a complexing agent [7,8], etc. It is well known that H_2O_2 being used in all copper CMP slurries as an oxidizer in order to make stable oxide passive layer on the copper surface. However, it was observed that copper oxide layer by H_2O_2 is not a passive film by (a) potentiodynamic measurements (no level off region) and (b) surface morphology from SEM. In previous studies, we have reported that a passive layer is necessary to obtain high removal rate as well as low roughness in tungsten CMP [10,11]. In this paper, we have investigated iodate and iodine based slurry chemistries for copper CMP with the goal of forming a copper-compound passive layer instead of oxide layer on the surface in order to obtain high removal rate with good planarity and without any surface damage, operating in intermediate pH ranges. We have used potassium iodate (KIO_3) and iodine (I_2) solution as a copper oxidizer, benzotriazole (BTA) as an inhibitor of copper corrosion, and potassium iodide (KI) [12,13] as an additive to supply iodide ions. In addition, the CMP results of H_2O_2 based slurry are shown to compare our results using iodate and iodine based slurry chemistries. The characteristics of these slurry chemistries have been measured by d.c. electrochemical measurements (potentiodynamic measurements) and inductively coupled plasma-atomic emission spectroscopy (ICP-AES) measurements. Also, scanning electron microscopy (SEM) with energy dispersion spectroscopy (EDS) was used for the characterization of surface layer morphologies and compositions.

EXPERIMENT

The slurry chemistries used in these experiments are presented in Table 1. In addition, slurries containing 0.01N I_2 were also used. The slurry pH was adjusted using HNO_3 and NaOH. In order to characterize the reaction between slurries and the copper surface, potentiodynamic polarization measurements were performed with a CH Instrument® Electrochemical Workstation Model 660 at a scan rate of 1 mV/s. Three electrode cell was used. High Purity copper foils (99.9985 %) purchased from Aldrich were used as the working electrode. These foils were cut into 1 cm^2 pieces, pre-polished mechanically and rinsed by acetone, methanol and DI water to remove any oxide and organic materials from surface before experiments. A saturated calomel reference electrode (SCE) and a graphite counter electrode were used for all the experiments. After immersion of Copper samples, they were immediately cathodically polarized (-0.7 V vs. SCE) for 2 min to prevent any oxide formation on the surface before potentiodynamic polarization measurements and ICP-AES measurements. To determine how protective the passive layer is, dissolution tests were carried out by placing the copper samples (1 cm^2) in 50 ml electrolytes for 30 min with vigorous stirring. Dissolved copper in electrolytes was determined using ICP-AES because the dissolution rate determined from the potentiodynamic curve is not accurate. In order to determine the chemical composition and morphology of the copper surface after potentiodynamic measurements, EDS and XRD were used.

Cu wafers were used for CMP experiments. These consisted of physical vapor deposited 150 nm copper seed layer, followed by electroplated 1.1 μm copper films with an average surface roughness (Rms) 19.5-nm on 25-nm Ta/SiO$_2$/Si substrates. 1.5 inch by 1.5 inch Samples were cleaved from 8-inch wafers. A Struers® Rotopol 31 polisher was used for polishing experiments, along with IC-1000/SUBA IV stacked pads. The pad was hand-conditioned with a grid-abrade diamond pad conditioner at the beginning and between each polishing run. The downward pressure applied on the samples was 2.9 psi. The head and the pad speed were 150 rpm, respectively. The flow rate of slurries was 100 ml/min. The slurries were prepared with Nanophase® Nanotek 30nm γ-alumina particles (manufacturer specification) with the slurry chemistries mentioned in Table 1. The solid concentration was 3% by weight for all measurements. The removal rates were determined by measuring the thickness of samples before and after polishing, using the four-point probe method.

RESULTS AND DISCUSSION

The potentiodynamic measurements carried out in the different solution in pH 4, 6 and 8 are shown in Fig. 1. It is well known that the Tafel curve indicates surface film formation and whether the surface film is passive or porous [2]. It can be observed that the anodic electrochemical behavior of copper in the polarization curves indicates the formation of a protective passive layer on the surface in all electrolytes from OCP value changes and level-off region in potentiodynamic curves. These surface layers can prevent further oxidation of copper, which is confirmed by measuring copper dissolution rate from ICP-AES measurements (Table 1). From Fig. 1 (a), it was seen that 0.1M KIO$_3$ slurry had passive region of copper in pH 6 and 8 due to the formation of oxide layer on the surface. This result corresponds to the Pourbaix diagram for copper [9]. As BTA is a well known inhibitor of copper corrosion, the potentiodynamic curve of 0.1M KIO$_3$ and 10^{-3}M BTA solution showed Cu-BTA layer formed on the surface, conformed by XPS results [14]. Also, lower current density and lower dissolution

Table. 1. The slurry chemistries used in experiments, dissolution rates of copper calculated from concentration of copper (ppb(w/v))

Slurry chemistry	pH	Dissolution rate (Å/min)	Removal rate without abrasives (Å/min)
0.1M KIO$_3$	4	28.5	207
	6	2.0	61
	8	0.7	85
0.1M KIO$_3$ and 10^{-2}M KI	6.5	0.2	75
	8	0.2	85
0.1M KIO$_3$ and 10^{-3}M BTA	4	0.04	1703
	6	0.03	187
	8	0.04	39
10% H$_2$O$_2$	4	1.1	
	6	0.4	
	8	0.4	

rate indicate that the Cu-BTA layer formed uniformly and was protective. On the other hand, since copper is in the active region in pH 4 in water system, it was expected that high removal rate could be obtained in CMP using 0.1M KIO$_3$ solution in pH 4. However, it had low removal rate (Fig. 2) and localized corrosion areas were observed on the surface after CMP. This result implied that CuI was formed partially by oxidation reaction, which was reported in ref [15] and confirmed by XPS results, and easily removed. The rest of the surface is probably oxide layer. Therefore, if KI is put into KIO$_3$ solution, it might supply sufficient iodide ions enabling the CuI compound layer to form uniformly on the surface, thus to obtain higher removal rate. Fig. 1 (c) shows the potentiodynamic curve of this solution. OCP value change and level-off region could be seen. In addition, this solution had lower copper dissolution rate (Table 1) and higher removal rate (Fig.2) than the solution containing only KIO$_3$. To enable faster formation of CuI layer, iodine (I$_2$) based solution was used because I$_2$ is oxidizer for copper and forms CuI. The reduction reactions of iodate and iodine are as follows:

$$IO_3^- + 6H^+ + 6e \leftrightarrow I^- + 3H_2O \qquad E^\circ = 1.085 \text{ V}$$
$$I2 + 2e \leftrightarrow 2I^- \qquad E^\circ = 0.5355 \text{ V}$$

Since KIO$_3$ consumes six hydrogen ions and six electrons and produces one iodine ion, additional iodine ions are required to form continuous CuI layer uniformly on the copper surface. Therefore, CuI layer formation in KIO$_3$ and KI solution is limited by mass-transfer control between the solution and the surface. However, iodine consumes two electrons and produces two iodine ions by reaction with copper. Therefore, a continuous CuI layer can be a product of this reaction. Then, the formation of CuI layer is controlled by surface reaction. From the Fig 1. (d), it can be seen that OCP was changed, which can be said surface layer was formed. However, there was no level-off region, which means that layers did not effectively passiving the copper surface, except 0.01N I$_2$ solution in pH 4. The narrow range of OCP in pH 4 indicates the passivation of the copper corrosion. H$_2$O$_2$ solution also showed same results (Fig. 1 (e)). These results can be explained by forming porous layers. In order to characterize the morphology and the composition of the surface layer, SEM with EDS was used to characterize the copper samples after they were dipped for 5 sec and 10 min with vigorous stirring, and after potentiodynamic measurements. As seen in Fig. 3, CuI layer was formed on the surface in all iodate and iodine based slurries, and copper oxide layer was formed on all H$_2$O$_2$ based slurries. However, the surface layer formation rate was different in all slurries. In

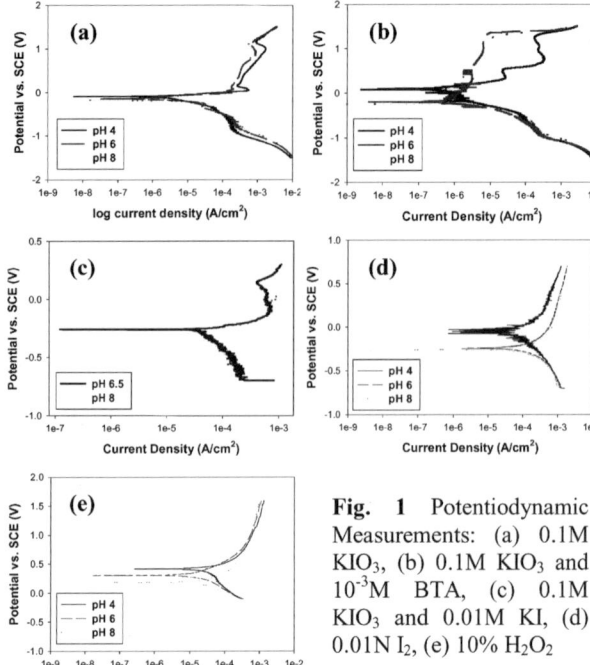

Fig. 1 Potentiodynamic Measurements: (a) 0.1M KIO$_3$, (b) 0.1M KIO$_3$ and 10^{-3}M BTA, (c) 0.1M KIO$_3$ and 0.01M KI, (d) 0.01N I$_2$, (e) 10% H$_2$O$_2$

case of KIO$_3$ and KI slurries, CuI (characterized by EDS) formed in islands after 10 min dipping in static OCP (OCP of potentiodynamic measurements) and CuI layer covered the copper surface partially after 5 seconds dipping in dynamic OCP (OCP during CMP) in Fig 3 (a) and (b). As mentioned earlier, the formation of CuI layer in the I$_2$ solution is dependent on the surface reaction rate, whereas diffusion of I$^-$ from solution to the surface may limit the CuI reaction rate in KIO$_3$ and KI solution. As seen in Fig. 3 (c) and (d), CuI layer formed uniformly on the surface after 5 seconds in static OCP. Also, it was observed that as pH increased, the coverage of CuI on the surface was decreased due to the formation of oxide in high pH. Despite CuI formation on the whole surface area, dissolution rate was high and no level-off region seen in Fig. 1 (d) because CuI layer

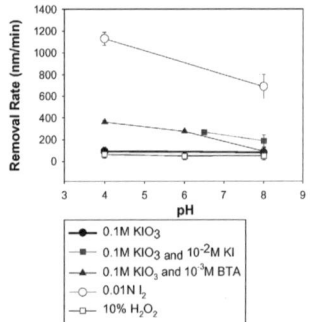

Fig. 2 Removal rates as a function of pH in different slurries.

could not passivate effectively from copper corrosion. But, in pH 4 CuI layer behave like aluminum oxide layer (aluminum has broad range of OCP). From these results, it can be inferred that CuI monolayer formed like continuous film, then as it became thicker by applying potential, films were broken and formed particle-like due to the severe lattice mismatch between copper and copper iodide. The surface after potentiodynamic measurements in H_2O_2 solution was also investigated. No surface morphology change was seen after dipping test by SEM, so a higher driving force for oxidation of copper was given by applying potential for long time (about 20 min from OCP to 1.5 V). From Fig 3 (e) and (f), needle shaped oxide layer (probably Cu_2O, because its color on the surface was violet red), confirmed by EDS, could be seen. These pictures indicate that copper oxide layer is porous layer and the formation of oxide layer in H_2O_2 solution is controlled by slow reaction.

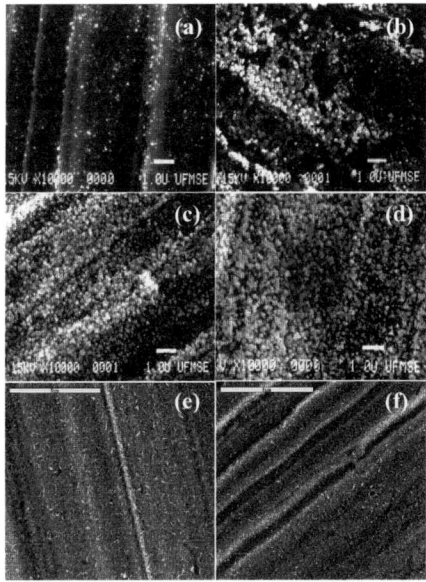

Fig. 3 Copper Surface Morphology by SEM: (a) 0.1M KIO_3 and 0.01M KI in pH 6.5, 10 min dipping, (b) 0.1M KIO_3 and 0.01M KI in pH 6.5, 5 sec dipping in dynamic OCP, (c) 0.01N I_2 in pH 4, 5 sec dipping, (d) 0.01N I_2 in pH 6, 5 sec dipping, (e) after potentiodynamic measurement in 10% H_2O_2 in pH 4, (f) after potentiodynamic measurement in 10% H_2O_2 in pH 6

CONCLUSION

Iodate and iodine slurry chemistries for copper CMP have been investigated. They are based on forming Cu-BTA and CuI layers on the copper surface by adding BTA and KI into the solution containing KIO_3, respectively, and the faster formation of CuI layer in I_2 based slurries. The experimental results show the formation of CuI layer on the surface as observed by the anodic electrochemical behavior of copper and the surface analyses of EDS. From potentiodynamic measurements and SEM results, the solution containing KIO_3 and KI can form CuI layer and higher removal rate (268 ± 16 nm/min) in pH 6.5 can be achieved than only KIO_3 due to the formation of CuI layer instead of oxide layer. But its reaction rate was slow. It was determined that 0.01N I_2 were very effective in forming a passive layer in pH 4, and the highest removal rate with this slurries could be achieved. In addition, It was seen that CuI was uniformly formed very fast in I_2 solution, since CuI is the product of the oxidation reaction of copper with I_2 solution, whereas diffusion of I^- may limit the CuI reaction rate in KIO_3 and KI solution. Therefore, the formation of CuI layer on the copper surface is more effective in CMP of copper, since CuI passive layers can be removed with lower pressure and lower concentration of abrasives than copper oxide layers, scratches problems could be diminished

ACKNOWLEDGEMENT

The authors would like to acknowledge the financial support of the Engineering Research Center (ERC) for Particle Science and Technology at the University of Florida, the National Science Foundation (NSF) grant #EEC-94-02989, and the Industrial Partners of the ERC

REFERENCE

1. F. B. Kaufman, D. B. Thompson, R. E. Broadie, M . A. Jaso, W. L. Guthrie, D. J. Pearson and M. B. Small, J. Electrochem. Soc., **138**, 3460 (1991)
2. J. M. Steigerwald, S. P. Murarka and R. J. Gutmann, Chemical mechanical planarization of microelectronic materials, John Wiley and Sons, New York (1997)
3. R. Carpio, J. Farkas and R. Jairath, Thin Solid Films, **266**, 238 (1995)
4. Q. Luo, S. Ramarajan and S. V. Babu, Thin Solid Films, **335**, 160 (1998)
5. M. Rutten, P. Feeney, R. Cheek, and W. Landers, Semiconductor Int., 123 (Aug. 1995)
6. Z. Stavreva, D. Zeidler, M. Plotner, and K. Drescher. Appl. Surf. Sci., **91**, 192 (1995)
7. J. M. Steigerwald, S. P. Murerka, R. J. Gutmann, and D. J. Duquette, J. Vac. Sci. Technol., **13B**, 2215 (1995)
8. J. M. Steigerwald, S. P. Murerka, R. J. Gutmann, and D. J. Duquette, Mater. Chem. Phys. **41**, 217 (1995)
9. M. Pourbaix, Atlas of Electrochemical Equilibria in Aqueous Solution, NACE, Houston, TX 1974
10. M. Bielmann, U. Mahajan, and R. K. Singh, Electrochem. and Solid-state lett., **2**, 401 (1999)
11. M. Bielmann, U. Mahajan, R. K. Singh, D. O. Shah, and B. J. Palla, Electrochem. and Solid-state lett., **2**, 148 (1999)
12. Y. C. Wu, P. Zhang, H. W. Pickering and D. L. Allara, J. Electrochem. Soc., **140**, 2791 (1993)
13. D. P. Schweinsberg, S. E. Bottle, and V. Otieno-Alego, J. Appl. Electrochem., **27**, 161 (1997)
14. S-M Lee, U. Mahajan, Z. Chen and R. K. Singh, Proceeding of the Electrochemical Society, 1999 Fall.
15. D. Tromans and J. Silva, J. Electrochem. Soc., **143**, 458 (1996)

Process Integration and
Manufacturability

Mat. Res. Soc. Symp. Vol. 613 © 2000 Materials Research Society

Technique of surface control with the Electrolyzed D.I.water for post CMP cleaning

Mitsuhiko Shirakashi*, Kenya Itoh*, Ichiro Katakabe*, Masayuki Kamezawa*,
Sachiko Kihara*, Manabu Tsujimura*, Takayuki Saitoh**, Kaoru Yamada**,
Naoto Miyashita***, Masako Kodera***, Yoshitaka Matsui***

* Precision Machinery Group, Ebara Corporation
   4-2-1,Honfujisawa, Fujisawa-shi 251-8502, Kanagawa, Japan
** Center of Technology Development, Ebara Research CO., Ltd.
   4-2-1,Honfujisawa, Fujisawa-shi 251-8502, Kanagawa, Japan
*** Manufacturing Engineering Center, Toshiba Corporation Semiconductor Company
   8,Shinsugita-cho, Isogo-ku, Yokohama 235-8522, Kanagawa, Japan

ABSTRACT

Chemical mechanical planarization (CMP) has been widely used for planarization of ILD, STI, plug and wiring processes. Wafer has several surfaces of materials, such as wiring materials, barrier materials, dielectric materials etc., that must be cleaned at the same time. In post metal CMP cleaning processes, in addition to cleaning several surfaces, it is very important that the oxidization level of metal materials, such as wiring, is held and controlled to maintain its resistance. Especially copper, that is began to use for wiring, is very easy to be oxidized. We have confirmed that the Electrolyzed D.I.water is effective in post Cu CMP cleaning for controlling the surface condition of Cu during cleaning and leaving a robust surface after CMP. We describe the Electrolyzed D.I.water system and present some result of analysis of Cu surface by treated with the Electrolyzed D.I.water.

INTRODUCTION

Wet processes such as CMP and electrochemical deposition (ECD) have recently received a tremendous amount of attention in the semiconductor industry in accordance with the progressive down sizing of the technical node on semiconductor devices. The integration of these processes into the semiconductor industry is primarily for the superior planarity (CMP) or gap filling (ECD) capabilities compared to the existing dry processes. The main reason why such wet processes have not been adopted into semiconductor processing in the past has been because wet processing has been considered "dirty" both from a process perspective and from a tool design perspective. The recent acceptance of wet processes and tools into semiconductor processing facilities is due to the recognition that such processes can be done at lower temperatures ($<100^{\circ}C$) and the introduction of the "dry-in dry-out" concept allows for cleaning at these low temperatures prior to future high temperature dry processes. This "dry-in dry-out" concept, which is now accepted as the industry standard, uses built-in cleaning and drying technologies that are critical to these wet processes.

The RCA cleaning approach which uses hydrochloric acid, ammonia, sulfuric acid, hydrofluoric acid etc. has been widely adopted so far. However, as the industry moves to larger wafer sizes (200 mm and 300 mm), it is very important to minimize the cost of chemical consumption and waste treatment.. The electrolyzed D.I. water system and processes were developed in accordance with this cost and environmental requirement. This report focuses on control of the oxidization level of the copper metal wiring surface after CMP using Electrolyzed D.I.water. Although the main purpose of these cleaning experiments is to remove foreign materials, control of the surface condition after cleaning is also important for the next process step. The superiority of electrolyzed water in semiconductor device cleaning over water containing dissolved gasses is demonstrated

EXPERIMENTAL

The cleaning mechanisms for the following experiments are defined as follows:

(1) Physical removal of foreign materials from surface.
(2) Control of the electrical potential and pH.
(3) Chemical etching of wafer surfaces.
(4) Removal of foreign materials in controlled chemical environment.

The principle of the Electrolyzed D.I.water is shown on Figure 1.

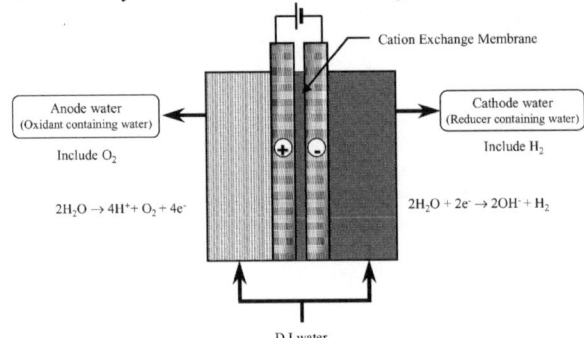

Figure 1 : Principle of generating the Electrolyzed D.I.water

The electrolyzing cell utilizes a cation exchange membrane between the anode and cathode separating the electrolyzed D.I.water in each compartment. Reactions occurring in the anode and cathode compartments are as follows:

Anode side: $2H_2O \rightarrow 4H^+ + O_2 + 4e^-$
Cathode side: $2H_2O + 2e^- \rightarrow 2OH^- + H_2$

In this reaction, oxidant containing solution (named anode water) is generated and reductant containing solution (named cathode water) is generated in the cathode compartment.

Figure 2 shows a schematic of the flow in the electrolyzed D.I.water supply system.

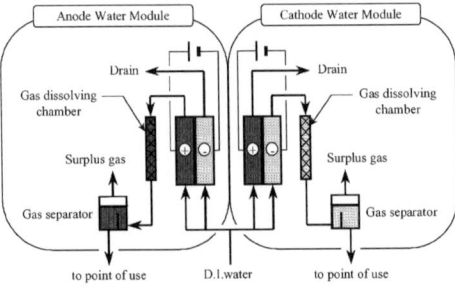

Figure 2 : Flow of the Electrolyzed D.I.water supply system

The system has two electrolyzing cells to allow the user to generate Anode and Cathode water simultaneously or independently. Gas dissolving chambers are mounted in sequence with the cells in order to enhance thorough dissolution of the generated gasses into solution. Surplus gasses are extracted from the anode and cathode water using gas separators before, each is supplied to the point of use.

Figure 3 shows a picture of the electrolyzed D.I.water supply system used in this study.

Figure 3 :
Appearance of
the Electrolyzed D.I.water supply unit

Figure 4 shows a typical experimental procedure. 200 mm diameter blanket wafers with electroplated copper were used. The wafer surface is initialized by rinsing with D.I.water followed by light etching of the oxidized surface with DHF. The cleaning steps were done using an ultrasonic nozzle with DI water as the control, or the appropriate solution. The effect of anode water is tested compared with oxidizing solutions containing dissolved gases (O_3 water and O_2 water). Reference wafer is cleaned only by DI water. X-ray Photoelectron Spectroscopy (XPS) was used to obtain the composition ratio of Cu, Cu oxide and Cu hydroxide by analysis of the Cu2p electron, Cu Auger electron and O1s electron.

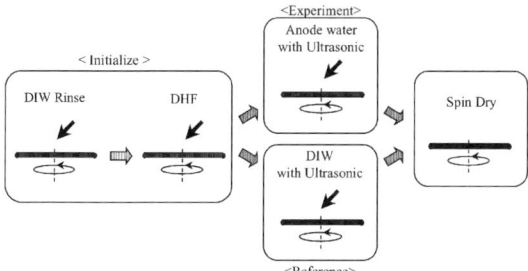

Figure 4 : Procedure of experiment

RESULTS AND DISCUSSION

In order to confirm the surface control by Anode water, the level of copper oxidization was analyzed by XPS. After four kinds of cleaning treatments, (1)D.I.water (2)O_3 water (3)O_2 water (4)Anode water, XPS results were obtained and compared on wafers 24 hours and 2 weeks after cleaning.

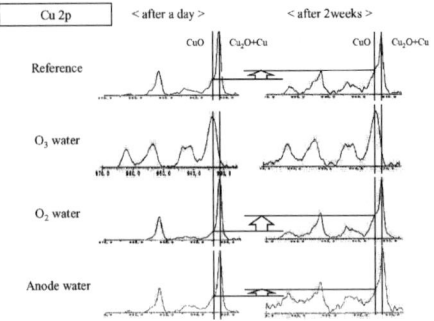

Figure 5 : Results of Cu2p spectrum analyses by XPS

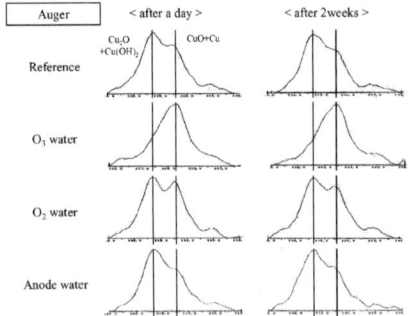

Figure 6 : Results of Cu Auger spectrum analyses by XPS

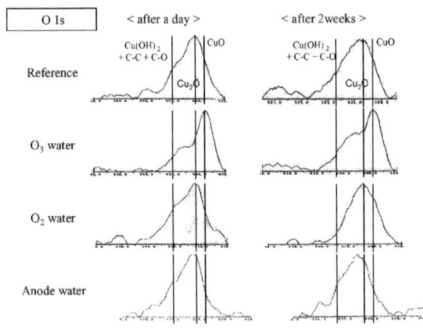

Figure 7 : Results of O1s spectrum analyses by XPS

The results of Cu2p spectrum analyses by XPS are shown on Figure-5. The results of Cu Auger spectrum analyses are shown on Figure-6. The results of O1s spectrum analyses are shown on Figure-7.

<u>24 hours after cleaning treatments:</u>

CuO peak of O_3 water treatment is biggest than peaks of other three cases in Figures-5, -6, -7. CuO peak of O_2 water is higher than other treatments in Figure-6. CuO peak of DI water is higher than O_2 water and Anode water cases in Figure-7. The data shows that copper oxidization levels were in the following order:

O_3 water > DI water > O_2 water > anode water.

with O_3 water exhibiting the most oxidation and anode water showing the least amount of oxidation.

<u>2 weeks after cleaning treatments:</u>

In case of O_3 water, oxidization level is still remained high even after 2 weeks.
CuO levels of all treatment cases except O_3 water treatment are almost same as shown on Figure-5. Figure-7 shows that oxidization continued in case of O_2 water because CuO peak is increased and $Cu(OH)_2$ peak is decreased.
The data suggests that in the case of wafers treated with O_3 water oxidation is the highest and remains stable, whereas in the case of wafers treated with O_2 water oxidation is not as high as on wafers treated with O_3 water, but continues to grow as oxygen from the environment is able to penetrate the surface oxide.

Depth analysis of the oxygen content ratio is shown on Figure-8. The data compares wafers treated with O_2 water and anode water. Broken lines and solid lines show respectively oxygen ratio data both 24 hours and 2 weeks after treatments. The difference between broken lines and solid lines shows the change in the oxygen content ratio.

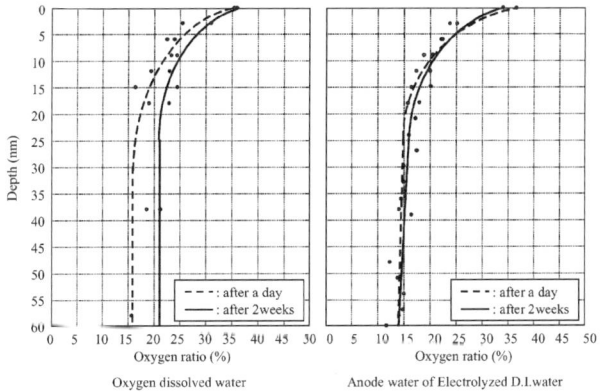

Figure 8 : Oxygen content in Cu

Oxygen ratio for wafers treated with O_2 water has clearly increased up to a depth of 60 nm, but the ratio for wafers treated with anode water is stable. It is postulated that the nature of the thin oxide film that is created by anode water treatment prevents oxygen penetration and prevents the progression of further oxidation.

CONCLUSION

The treatment of copper wafers with anode water has been demonstrated as a viable method to prevent post CMP oxidation of copper lines in semiconductor devices. Oxidization levels on copper surfaces after four kinds of treatments, (1) D.I.water, (2) O_3 water, (3)O_2 water, and (4)Anode water, were compared.

(1) CuO ratio on wafers treated with anode water was stable and low compared with the treatments by DI water, O_2 water and O_3 water even 2 weeks after treatment.
(2) Oxidization on wafers treated with anode water was shown to be stable and low even 2 weeks after treatment using depth analysis.

Based on the results obtained in this report, further studies on actual devices will be conducted.

REFERENCES

[1]N.Miyashita, Y.Mase, J.Takayasu, Y.Minami, M.Kodera, M.Abe, T.Izumi: "Mechanism of a New Post CMP Cleaning for Trench Isolation Process", Materials Research Society 1999 Spring Meeting, San Francisco, CA, April 1999.
[2]D.Briggs, M.P.Seah: "Practical Surface Analysis by Auger and X-ray Photoelectron Spectroscopy", John Wiley & Sons Ltd.
[3]The surface science society of Japan: "Method of X-ray Photoelectron Spectroscopy", Maruzen

Mat. Res. Soc. Symp. Vol. 613 © 2000 Materials Research Society

REMOVAL RATE, UNIFORMITY AND DEFECTIVITY STUDIES OF CHEMICAL MECHANICAL POLISHING OF BPSG FILMS

Benjamin A. Bonner, Boris Fishkin, Jeffrey David, Chad Garretson, and Thomas H. Osterheld
CMP Division, Applied Materials
3111 Coronado Drive, Santa Clara, CA 95054

ABSTRACT

Wafers where deposited with BPSG films having phosphorus concentration varying from 3.65 to 6.25% and boron concentration varying from 4 to 5.7%. These wafers were polished using CMP and the rates were found to depend on dopant concentrations. A fit to the data indicated that removal rates were more than 3 times as sensitive to boron concentration compared to phosphorus concentration. For a constant phosphorus concentration of 5%, each percent increase in boron increases CMP removal rate by 340 Å/min. For a constant boron concentration of 5%, each percent increase in phosphorus increases CMP removal rate by 96 Å/min.

INTRODUCTION

Borophosphorosilicate glass (BPSG) is currently a film of choice as pre-metal dielectric [1-4]. The addition of phosphorous to silicate films may lower the migration of alkali ions, while boron addition lowers the glass transition temperature of the film allowing it to flow at lower temperatures to give better local planarization [1,2,5]. The move toward sub-0.25 micron line width requires global planarization to achieve good depth of focus. This global planarization can be achieved by chemical mechanical polishing (CMP).

EXPERIMENTAL DETAILS

Polishing experiments were performed on an Applied Materials Mirra® CMP system equipped with the Oxide Plus hardware package. All experiments used SS-12 slurry from Cabot Corporation and IC1010 pads from Rodel.

Oxide thickness was measured on a Thermawave Optiprobe 3260. Oxide thickness was measured using a standard 49-point contour map having a 5 mm edge exclusion. Removal rates were calculated from average thickness using the 49-point contour map. Within-wafer-nonuniformity (WIWNU) was expressed as a percentage and calculated using standard deviation of removal divided by average removal.

BPSG wafers were deposited in an Applied Materials Giga-Fill™ SACVD chamber. Boron concentration was varied between 4 and 5.7% while phosphorus concentration was varied between 3.65 and 6.25%. Wafers were annealed by standard methods after deposition.

RESULTS AND DISCUSSION

Doping an oxide film with boron or phosphorus increases oxide removal rate significantly [2,5-7]. Under the baseline BPSG polish conditions used for these experiments, thermal oxide (undoped) has a removal rate of 1590 Å/min (standard ILD polish conditions yield a much higher removal rate). In contrast, all of the oxide films doped with boron and phosphorus had removal rates of more than 4000 Å /min. To better understand this behavior, a series of experiments were performed on wafers deposited with varying amounts of boron and phosphorus.

For the first series of experiments, wafers were prepared with 5 different phosphorus concentrations by varying the flow rate of the phosphorus precursor during deposition. The boron precursor flow-rate was held constant during the deposition of conditions 1-5. The weight percent of each dopant was measured after deposition using X-ray fluorescence and the resulting dopant concentrations are summarized in Table I. Table I shows that phosphorus concentration is varied from 3.65% by weight to 6.25% by weight. Some variation in boron concentration is also observed even though the precursor is held at a constant flow rate.

Table I. Boron and Phosphorus concentrations for Conditions 1-5

Condition	B wt.%	P wt.%
1	4.80	3.65
2	4.90	4.40
3	5.00	5.10
4	5.25	5.65
5	5.25	6.25

For the second series of experiments, wafers were prepared with 5 different boron concentrations by varying the flow rate of the boron precursor during deposition. The phosphorus precursor flow-rate was held constant during the deposition of conditions 6-10. The weight percent of each dopant was measured after deposition using X-ray fluorescence and the resulting dopant concentrations are summarized in Table II. Table II shows that boron concentration is varied from 4.0% by weight to 5.7% by weight. Some variation in phosphorus concentration is also observed even though the precursor is held at a constant flow rate.

Table II. Boron and Phosphorus concentrations for Conditions 6-10

Condition	B wt.%	P wt.%
6	4.00	5.60
7	4.50	5.35
8	5.00	5.10
9	5.35	4.95
10	5.70	4.80

Two wafers were polished for each dopant condition using a standard BPSG polishing process. Average removal rate and WIWNU for the polish process were obtained using pre and post-polish measurements. Results of these experiments are summarized in Table III.

Inspection of conditions 1 through 5 in Table III clearly shows that removal rate increases as phosphorus concentration increases. For example, condition 1 with 3.65% phosphorus has a rate of 4154 Å/min while condition 5 with 6.25% phosphorus has a higher removal rate of 4599 Å/min. Similarly, inspection of conditions 6-10 shows that removal rate increases as boron concentration increases. For example, condition 6 with 4.0% boron has a rate of 4202 Å/min while condition 10 with 5.7% boron has a higher removal rate of 4680 Å/min.

Table III. Removal rate and WIWNU for polished BPSG wafers

Condition	B (wt%)	P (wt%)	Removal Rate (Å/min)	WIWNU (%)
1	4.80	3.65	4154	3.44
2	4.90	4.40	4303	3.32
3	5.00	5.10	4325	2.83
4	5.25	5.65	4460	2.99
5	5.25	6.25	4599	3.59
6	4.00	5.60	4202	2.68
7	4.50	5.35	4255	3.97
8	5.00	5.10	4325	2.83
9	5.35	4.95	4581	3.36
10	5.70	4.80	4680	2.70

Removal rate dependence on varying boron and phosphorus concentration is plotted in Figure 1. The best-fit lines for both sets of conditions are also provided in the graph because the dependence on P or B concentration appears reasonably linear in this range. The slope for the fit to varying phosphorus concentration is 160 (Å/min)/(% P) with an R^2 of 0.95 while the slope for the fit to the varying boron concentration is 290 (Å/min)/(% B) with an R^2 of 0.885.

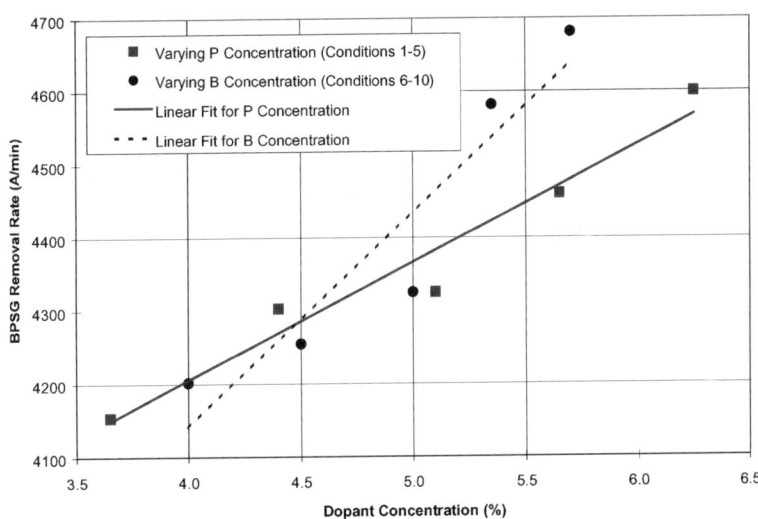

Figure 1. *CMP Removal rate dependence on phosphorus or boron concentration*

Inspection of Tables I and II shows that boron concentration was not constant for conditions 1-5 (designed to only vary phosphorus concentration) while phosphorus concentration was not constant for conditions 6-10 (designed to only vary boron concentration). At the same time, Figure 1 shows that removal rate depends on both phosphorus and boron concentration. It would be valuable to isolate the contribution of phosphorus and boron to removal rate.

Because the slopes are so different for rate dependence on boron vs. phosphorus concentration, one would not expect a dependence on total dopant concentration (sum of phosphorus and boron concentration). This conclusion is borne out by a plot of rate vs. total dopant concentration where conditions 1-5 are not collinear with conditions 6-10. Instead, a multiple regression analysis was performed to obtain the dependence on boron and phosphorus concentrations. This analysis resulted in Equation (1) having an R^2 of 0.83.

$$\text{Rate (Å/min)} = 335 \text{ (Å/min/\%B)} \times \text{(B Conc)} + 100 \text{ (Å/min/\%P)} \times \text{(P Conc)} + 2235(\text{Å/min}) \quad (1)$$

The ratio of slopes from this equation is 335/100=3.35. The ratio of slopes indicates that doping with boron is 3.35 times as effective as phosphorus in increasing removal rate. An effective dopant concentration can be calculated by adding 3.35 times the boron concentration to the phosphorus concentration. A plot of removal rate dependence on this effective dopant concentration is provided in Figure 2. The best-fit line for all of the data (conditions 1-5 and conditions 6-10) is also provided. This fit has an R^2 of 0.87 indicating a reasonably valid correlation.

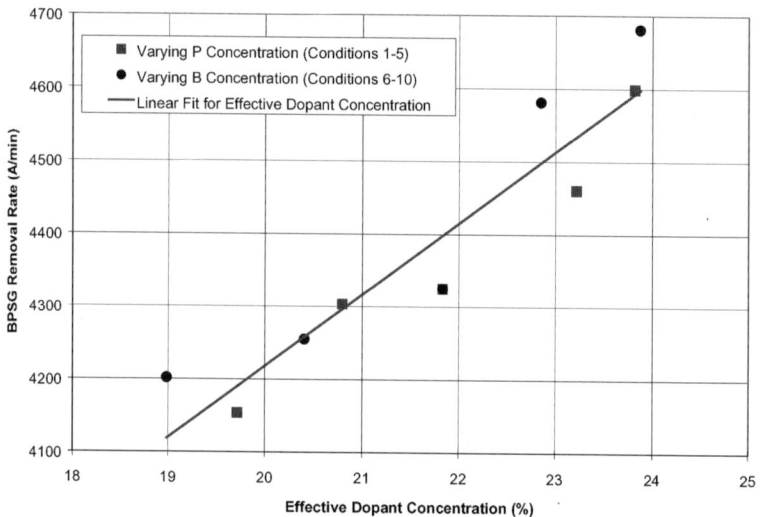

Figure 2. *CMP Removal rate dependence on effective dopant concentration (described in text)*

Equation (1) can be used to correct the removal rates in conditions 1-5 to constant boron concentration and the rates in conditions 6-12 to constant phosphorus concentration. The mid-point of 5% was selected in both cases. Table IV provides the results of this calculation.

Corrected removal rate dependence on varying boron and phosphorus concentration is plotted in Figure 3. The best-fit lines for both sets of conditions are also provided in the graph. The slope for the fit to varying phosphorus concentration is 96 (Å/min)/(% P) with an R^2 of 0.86 while the slope for the fit to the varying boron concentration is 340 (Å/min)/(% B) with an R^2 of 0.915. For a constant boron concentration of 5%, each percent increase of phosphorus increases the film removal rate by 96 Å/min. For a constant phosphorus concentration of 5%, each percent increase in boron concentration increases the film removal rate by 340 Å/min.

Table IV. Removal rate corrected to constant B or P concentration for polished BPSG wafers

Condition	B (wt%)	P (wt%)	Corrected B (wt%)	Corrected P (wt%)	Removal Rate (Å/min)	Corrected Removal Rate (Å/min)
1	4.80	3.65	5.00		4154	4221
2	4.90	4.40	5.00		4303	4337
3	5.00	5.10	5.00		4325	4325
4	5.25	5.65	5.00		4460	4376
5	5.25	6.25	5.00		4599	4515
6	4.00	5.60		5.00	4202	4142
7	4.50	5.35		5.00	4255	4220
8	5.00	5.10		5.00	4325	4315
9	5.35	4.95		5.00	4581	4586
10	5.70	4.80		5.00	4680	4700

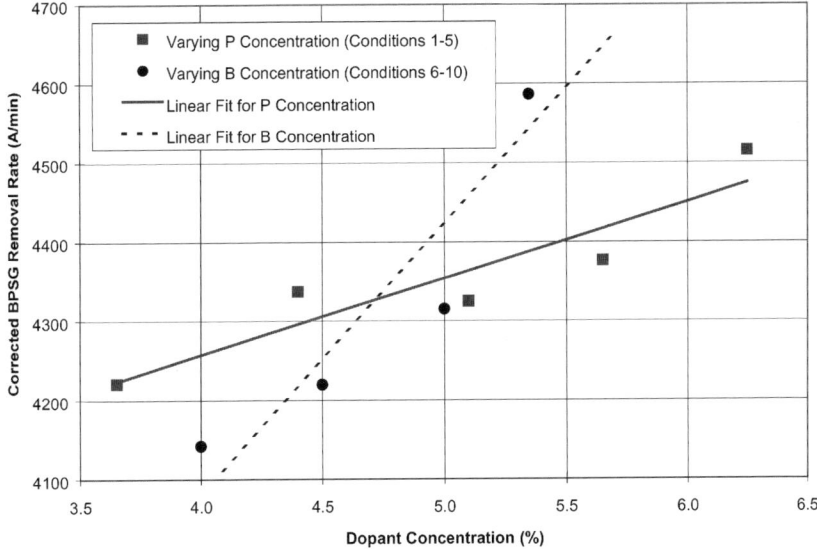

Figure 3. *Corrected CMP Removal rate dependence on phosphorus or boron concentration*

Several previous studies have investigated removal rate dependence on phosphorus concentration (no boron dopant) [1,2,5,6]. We are also aware of a study that tested sensitivity to boron concentration (no phosphorus dopant) [5]. Only one study showed non-normalized data and found approximately 150 Å/min increase in rate for each percent increase in phosphorus concentration [6]. This compares reasonably to 96 Å/min in this study. One would not necessarily expect good aggreement given that different process conditions and consumables were used in the two studies.

CONCLUSIONS

CMP of BPSG films showed that removal rates depend linearly on dopant concentrations. A fit to the data indicated that removal rates were more than 3 times as sensitive to boron concentration compared to phosphorus concentration. Coefficients from least-squared fits to the data were used to calculate a linear dependence on both boron and phosphorus concentration with an R^2 value of 0.87. These calculations were also used to isolate the removal rate dependence on both boron and phosphorus. For a constant phosphorus concentration of 5%, each percent increase in boron increases CMP removal rate by 340 Å/min. For a constant boron concentration of 5%, each percent increase in phosphorus increases CMP removal rate by 96 Å/min.

ACKNOWLEDGMENTS

The authors would like to acknowledge Paul Gee and Eugenia Liu of the SACVD group at Applied Materials for depositing the wafers used in these experiments and for numerous useful discussions.

REFERENCES

[1] S. J. Fang, S. Garza, H. Guo et al., "Optimization of the Chemical Mechanical Polishing Process for Premetal Dielectrices," Journal of the Electrochemical Society 147 (2), 682 (2000).

[2] C.-T. Ni, H. C. Chen, D. Huang et al., "A Study of CMP Slurry Chemistry Effect on BPSG Film for Advanced DRAM Application," presented at the Proceedings of Third International Chemical-Mechanical Planarization for the ULSI Multilevel Interconnection Conference (CMP-MIC), Santa Clara, CA, 1998 (unpublished).

[3] Joseph M. Steigerwald, Shyam P. Murarka, and Ronald J. Gutmann, Chemical Mechanical Planarization of Microelectronic Materials, 1 ed. (John Wiley & Sons, Inc., New York, 1997).

[4] M. Yoshimaru and H. Wakamatsu, "Microcrystal Growth on Borophosphosilicate Glass Film during High-Temperature Annealing," Journal of the Electrochemical Society 143 (2), 666 (1996).

[5] W.J. Schaffer, J. W. Westphal, H. W. Fry et al., "CMP Removal Rate and Nonuniformity of BPSG," presented at the Proceedings of First International Chemical-Mechanical Planarization for the ULSI Multilevel Interconnection Conference (CMP-MIC), Santa Clara, CA, 1996 (unpublished).

[6] S. Pennington and S. Luce, Proc. 9th VMIC 9 (92), 168 (1992).

[7] S. C. Sun, F. L. Yeh, and H. Z. Tien, Mat. Res. Soc. Symp. Proc. 337, 139 (1994).

Mat. Res. Soc. Symp. Vol. 613 © 2000 Materials Research Society

Using Wafer-Scale Patterns for CMP Analysis

Brian Lee[1], Terence Gan[1], Duane S. Boning[1], Jeffrey David[2], Benjamin A. Bonner[2], Peter McKeever[2], and Thomas H. Osterheld[2]
[1]Massachusetts Institute of Technology, Cambridge MA
[2]Applied Materials, Santa Clara, CA

ABSTRACT

A new set of wafer-scale patterns has been designed for analysis and modeling of key CMP effects. In particular, the goal of this work is to develop methods to characterize the planarization capability of a CMP process using simple measurements on wafer scale patterns. We examine means to pattern large trenches (e.g. 1 to 15 mm wide and 15 mm tall) or circles across 4" and 8" wafers, and present oxide polish results using both stacked and solo pads in conventional polish processes. We find that large separation (15 mm) between trenches enables cleaner measurement and analysis. Examination of oxide removal in the center of the trench as a function of trench width shows a saturation at a length comparable to the planarization length extracted from earlier studies of small-scale oxide patterns. Increase in polish pressure is observed to decrease this saturation point. Such wafer scale patterns may provide information on pad flexing limits in addition to planarization length, and promise to be useful in both patterned wafer CMP modeling and studies of wafer scale CMP dependencies such as nanotopography.

INTRODUCTION

Current techniques for characterizing CMP typically involve patterning test dies onto a wafer, running polish experiments, and analyzing measurements to obtain characterization parameters for the process [1]. A key parameter known as planarization length is typically used to describe the length scale over which feature-induced pattern density on the wafer affects the polishing at a particular point on the die. By adding feature-scale step-height considerations to planarization length-based density evaluation, accurate models of post-CMP oxide thickness (with ~100 Angstrom error) have been demonstrated for conventional stacked pads and processes [2].

An alternative approach using wafer-scale patterns has previously shown promise as a tool to study CMP pattern dependencies [3]. With an increased interest in harder polishing pads (to increase planarization length and reduce within-die variation), the planarization length is approaching the size of the typical die. In addition, harder pads may induce a "pad flexing limit" in which the contact of the pad in large "low" regions of the die is decreased. For these reasons, as well as interest in simplified measurement and analysis of planarization length, wafer-scale patterns for detailed CMP planarization characterization are explored further in this work.

In the next section, we describe new mask designs for wafer-scale patterns. Key issues include the range of trench (or circle) sizes that should be included, as well as the separation between them on the wafer. In addition, we describe two patterning methods used here, including traditional mask plates and acetate-based masks. The following section then summarizes the sets of wafer fabrication and polishing experiments conducted, followed by analysis and discussion of the experimental results. The relationship between the observed results and previous pattern density or contact wear models is briefly considered. Finally, we offer conclusions and suggestions for further work.

MASK DESIGN

Three sets of wafer-scale patterns are described here. Two of the designs are used for patterning 4-inch wafers, and the third design is used to pattern 8-inch wafers. The guiding principle behind the design of the masks is to fabricate structures of various sizes separated by relatively large distances to reduce interaction between these structures. The use of wafer scale patterns enables much larger structures and separations to be fabricated than is usually possible with conventional die patterns.

The layouts of the three patterns are summarized in Figure 1, where the width or diameter and relative positions of the trenches or circles are shown. The first pattern (Pattern A) is implemented as a standard quartz mask for use in a Karl Seuss contact aligner. The total pattern size is 70 mm x 70 mm, consisting of rectangular trenches 8 mm in height, ranging in width from 20 μm to 8 mm. The second pattern (Pattern B) is implemented using alternative transparency masks on 4" wafers. There are 30 circular trench structures in the layout, with 17 distinct widths ranging from 2 mm to 8 mm, and replicates of several of the structure sizes. These structures are separated from each other and from the edge of the wafer by at least 10 mm to reduce edge effects and interaction among structures. The structures are arranged in three concentric rings on the wafer. The final pattern (Pattern C) is implemented as a standard quartz mask for use in a contact aligner, with a total pattern size of 140 mm x 140 mm. It consists of 25 rectangular trench structures, each of which is 15 mm in height. There are 22 distinct widths, ranging in size from 20 μm to 15 mm, with replicates of the 1mm, 5mm, and 10mm trenches. These structures are separated from each other and from the edge of the wafer by at least 15 mm to further reduce edge effects and interaction among structures.

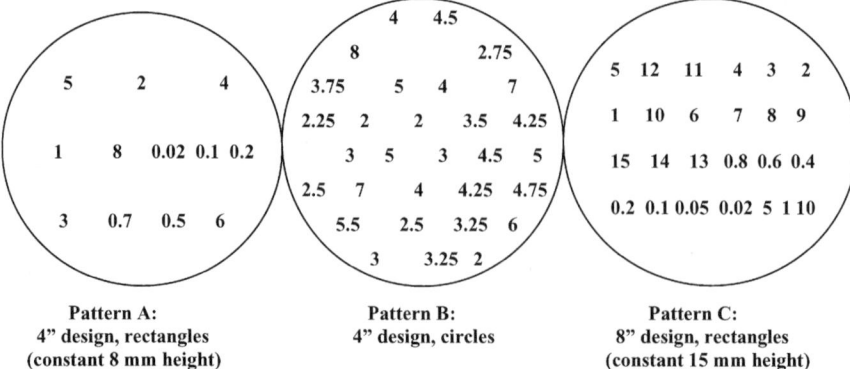

Pattern A:	Pattern B:	Pattern C:
4" design, rectangles	4" design, circles	8" design, rectangles
(constant 8 mm height)		(constant 15 mm height)

Figure 1. **Mask floor plan for wafer-scale masks (dimensions in mm).**
Positions of numbers indicate relative location of structures on the wafer.

In addition to consideration of structure size and separation issues, in this work we investigate an alternative means of patterning wafer-scale structures. Patterns A and C are implemented using traditional mask fabrication facilities. Pattern B, on the other hand, is implemented as a lithographic print on a translucent sheet of acetate (i.e., transparency), which is then used in a Karl Seuss contact aligner to pattern wafers. The lithographic print method is chosen to insure sufficient opaqueness of the pattern. This method results in poor structure edge resolution, and re-stricts the minimum structure size on the mask to 2 mm. For the study of large scale polishing behavior (large multi-mm structures), these limitations are not critical. The benefit of this method is that it enables inexpensive and rapid production of distinct wafer-scale patterns for characterization work.

EXPERIMENTAL DETAILS

Pattern A and Pattern B wafers are polished on a Strausbaugh 6EC CMP tool using a standard pad stack (IC1000/SubaIV) and slurry (Cabot SS-12). Pattern C wafers are polished on an Applied Materials Mirra™ tool using standard and harder pad stacks and standard slurry.

Pattern A wafers have 1.5 μm of CVD oxide, and are patterned and etched to produce 0.6 μm trenches using a wet etch process. The wafers are then polished under the following process conditions: 25 rpm table speed, 55 rpm quill speed, 3 psi down force, 1 psi back pressure, modifying the polish time to yield splits of three different thick-ness removals: 0.4 μm, 0.6 μm, and 0.8 μm.

Pattern B wafers have an initial CVD oxide of 1.5 μm, and are patterned (using the alternative mask method), and then etched to produce 0.7 μm trenches using a wet etch process. The wafers are then polished under the follow-ing process conditions: 25 rpm table speed, 15 rpm quill speed, 2.5 psi down force, 1 psi back pressure, modifying the polish time to yield splits of three different thickness removals: 0.3 μm, 0.5 μm, and 0.7 μm.

Pattern C wafers have trenches etched in silicon to an etch depth of 0.82 μm, with 1.5 μm of oxide then depos-ited across the wafer. Wafers are polished under three different process conditions (constant speed and varying pres-sures), but all splits are targeted towards the same amount of oxide removed (0.5 μm).

Thickness measurements are taken across the trenches using standard profilometry scans (in Pattern A), optical measurements (in Pattern B), or high-resolution optical measurements (in Pattern C).

EXPERIMENTAL RESULTS

One proposed method of using wafer-scale post-polish measurement data to characterize the CMP process is to analyze the amount of material removed in the center of trench structures as a function of the structure size. Wider trenches should result in more material removed, since the CMP pad can deform into the trenches to a greater degree. In this section, we examine the polish data from the trench structures of different sizes. Results indicate that wafer scale uniformity, as well as structure separation, are important considerations for wafer scale patterns. Given sufficient separation, trends in trench removal vs. structure size are clearly discernible.

Polish Depth and Structure Separation Guidelines

As expected, the amount of trench oxide removal increases as the etched structure size increases, as shown in Figure 4 for Patterns A and B. We see that the curve for Pattern B (which uses a non-traditional patterning step and circular trenches) exhibits the same general signature as the curve for Pattern A (which uses a traditional patterning step and rectangular trenches). Figure 4a shows that the general signature of the process does not change with the amount of material removed (approximately scaling with amount removed), provided that one does not polish past the depth of the trench itself. The curve for the 0.4 μm and 0.6 μm target removals show similar signatures, while the curve for the 0.8 μm removal exhibits a much noisier signature. We conjecture that wafer scale polish nonuniformities are exacerbated or that trench structures interact more strongly for large material removals, suggesting that moderate amounts of polish are best for wafer scale pattern studies.

Figure 4b shows the trend of trench area removal vs. structure size for the Pattern B wafers, where structures are replicated in three concentric rings on the wafer. We see that the general range of values and curve trends do not change depending on ring position. However, the structures nearer to the edge seem to exhibit a much noisier signal than the structures in the center of the wafer. Here again the 10 mm separation distance between structures and from the edge of the wafer may not be sufficient to block against structure interaction. The potential for wafer scale polish uniformity to affect structure polish also suggests that replication of trench structures is important in order to separate polish uniformity from trench size dependencies.

Pad and Process Impact on Trench Removal vs. Trench Width

Pattern C wafers, consisting of structures with larger 15 mm heights and separations, are polished using two different pads and three different polish processes; the trench removal vs. trench width plots are shown in Figure 5. The left side of Figure 5 shows results using a standard stacked pad, while the right side plots the results with a hard polishing pad. Three polishing processes in which only the head pressure varies are shown from top to bottom, where increasing pressure is from curves (a) to (c) and from curves (d) to (f).

Considering first the standard standard stacked pad and process of Figure 5(a), we see that the trench removal reaches a saturation point for trench sizes in the 5-6 mm range, at which point the trench removal is nearly constant (but less than in the surrounding unetched regions) for increasing trench widths. We also see that the trench removal appears to linearly decrease toward zero for smaller trench widths. As the polish pressure increases from Figure 5(a) to (c), we see that the saturation point becomes more pronounced and appears to occur at smaller trench widths. The difference between the trench and non-trench oxide removal for the largest trench sizes also decreases for the larger pressure processes.

Comparing these stacked pad results with the hard pad results shown in Figure 5(d) to (f), we see a dramatic difference in the saturation point for the harder pads. Indeed, for the lower pressure the saturation point is not reached for the largest 15 mm structure examined here. In the case of the highest pressure process, we see what may be a fairly sharp saturation at a trench width of 15 mm, compared to 4-5 mm for the stacked pad at the same pressure. For study of emerging hard pad CMP processes, these results suggest that wafer scale patterns with even larger structures and separation distances should be considered in the future.

DISCUSSION AND MODELS

In this section, we first consider the relationship between the observed data and a pattern density-based CMP model, and then briefly consider the above data from the perspective of a contact wear model.

Planarization Length and Pattern Density-Based Model

The "planarization length" is used to describe the ability of a CMP process to remove variation on a die [1]. The saturation point on the curve for the standard pad is in the 4-6 mm range, which is comparable to planarization lengths previously extracted for this process using pattern density test masks [4]. An idealized pattern density-based CMP model is considered here to show that this saturation point is the same as the planarization length parameter.

Figure 2 illustrates the trench polish problem using an idealized analysis, in which a simple square averaging window of size equal to the planarization length is used to calculate the "effective density" of raised topography around and within a trench. The effective density as calculated as the ratio of raised material to total area within some region defined by the planarization length. The effective density-based polish model assumes that raised or "up" areas of regions will polish as the blanket rate divided by the effective density of that region, and once the up structures are removed the region polishes at the blanket polish rate [1].

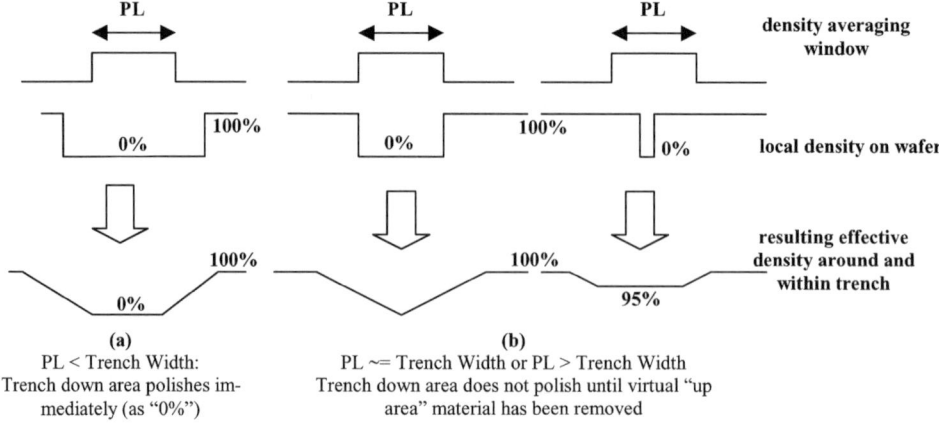

(a)	(b)
PL < Trench Width:	PL ~= Trench Width or PL > Trench Width
Trench down area polishes im-	Trench down area does not polish until virtual "up
mediately (as "0%")	area" material has been removed

Figure 2.   The trench polish problem using a planarization length (PL) and effective density analysis.

We see in Figure 2 that trenches with widths larger than the planarization length of the process have a trench central region that polishes as a "0% effective density area" surrounded by areas of monotonically increasing density. Regions more than half a planarization length away from the trench edge polish as 100% effective density regions (or at the blanket polish rate). In the "0% effective density area" the wafer should polish as the blanket rate. For trenches with lengths equal to or less than the planarization length, the trench central region will evaluate to a non-zero effective density, so that no down area (i.e. trench center) polish occurs until the up area material is removed. For the trench case where we have non-zero density points inside the trench, this would refer to virtual "up area" since points inside the trench have no real up area material. The hypothetical trench removal vs. trench width plot resulting from this effective density polish model is summarized in Figure 3.

While the effective density models can relate the saturation point in the plots of Figure 5 to the notion of planarization length, other aspects of the data are not well explained using such a model. First, as discussed earlier, at saturation the trench removal amount is less than the corresponding amount removed from outside of the trench, while the density model suggests they should both polish at the blanket rate (the dashed line in Figure 3(a)). Second, the data for both the standard and harder pads indicate that trench down area removal is zero only at very small trench sizes, while the density model predicts a substantial range of zero polish. Down-area polish before complete removal of local step height has been modeled [2; however, the step height at which this begins to occur (e.g. 2000 Angstroms) is less than the final trench height in this polish experiment. The polish results presented here may indicate that such step "contact heights" may be very large for such large structures.

Contact Wear Model

Since the effective density/planarization length approach was originally formulated consider polishing on the feature scale, an alternative method of analysis may be more suited for approaching the macroscopic wafer-scale polishing problem – that of considering pad/wafer contact mechanics [5,6]. This approach considers the physical interaction of the contact between an elastic pad and the wafer, and forms a relationship between the displacement of the pad and the pressure on the wafer. We implement a contact mechanics model similar to that in [5,6] and apply the model to the trench polish problem resulting in a hypothetical trench removal vs. trench width curve as shown in Figure 3(b). Using a contact mechanics argument, the pressure on the pad at the bottom of a trench is less than the pressure on the raised area, which directly translates to less material removed in the trench center than on the area outside the trench, thus explaining that characteristic of the observed data. The contact mechanics approach can also result in zero down polish for non-zero trench widths (i.e. there may be a width at which the pad does not contact the wafer). This aspect of the data observed in Figure 4 needs further exploration.

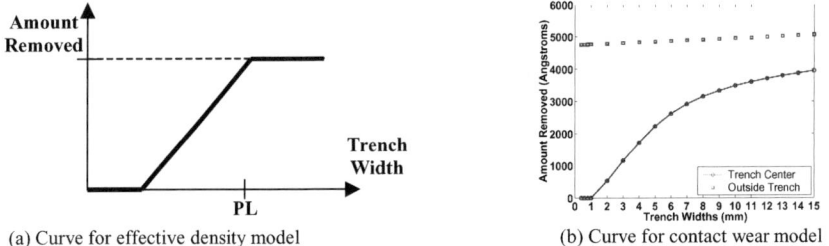

(a) Curve for effective density model   (b) Curve for contact wear model

Figure 3.   Hypothetical amount removed vs. trench width plots, for pattern density-based model (a), and contact wear model (b).

CONCLUSIONS

We have examined wafer-scale patterns as an alternative means for characterizing and modeling CMP processes. Polishing results suggest that such wafer scale patterns benefit from very large separations of structures from each other and the wafer edge to avoid interactions, particularly for harder pad or longer planarization length processes. Trench removal vs. trench width plots provide useful insight into the polish process, and can indicate a planarization length parameter. Such plots also reveal dependencies that merit further exploration (the difference between large trench and outside trench polish, the width at which trenches begin to polish) using contact mechanics and other modeling approaches. Future models of CMP pattern evolution may well require integration of both macroscopic consideration and feature scale behavior.

Wafer-scale pattern experiments and modeling may prove particularly useful in the study of "nanotopography" or "nanotopology" related to the nanometer-scale surface variations (occurring over mm length scales across the wafer) that may be present on bare silicon wafers [7]. It has been proposed that natural "random" nanotopography that occurs on an unpatterned raw silicon wafer can be approximated by using a fixed grid of randomly sized cylindrical posts on a wafer-scale pattern [8]. Polishing of such patterned films could lead to insights on how to model the CMP of natural nanotopography on wafers.

ACKNOWLEDGMENTS

The authors acknowledge the MIT Microsystems Technology Laboratories technicians for assistance in some of these experiments. We thank Peter Burke for discussions about his wafer-scale patterns, and Alvaro Maury from Lucent for early discussion of this approach. We also thank Michael Oliver from Rodel Inc. and Dale Hetherington at Sandia National Laboratories for discussions on this topic. This work has been supported in part by a DARPA subcontract with PDF Solutions.

REFERENCES

1. D. Ouma, *et al.*, "An Integrated Characterization and Modeling Methodology for CMP Dielectric Planarization," *International Interconnect Technology Conference*, San Francisco CA, June 1998.
2. T. Smith, *et al.*, "A CMP Model Combining Density and Time Dependencies," *CMP-MIC Conference*, Santa Clara, CA, Feb. 1999.
3. P. Burke, *et al.*, MRS, Oct. 1996.
4. R. Jin, *et al.*, "A Production-Proven Shallow Trench Isolation (STI) Solution Using Novel CMP Concepts," *CMP-MIC Conference*, Santa Clara, CA, Feb. 1999.
5. O.G. Chekina, *et al.*, "Wear-Contact Problems and Modeling of Chemical-Mechanical Polishing," *J. Elec. Soc.,* Vol 145, No. 6. June 1998.
6. T. Yoshida, "Three-Dimensional Chemical-Mechanical Polishing Process Model by BEM," *ECS*, Oct. 1999.
7. K.V. Ravi, "Wafer Flatness Requirements for Future Technologies," *Future Fab International*, Issue 7, pp. 207.
8. N. Poduje, *et al.*, "Nanotopology Effects in Chemical Mechanical Polishing," *SEMI-AWG Nanotopology Workshop*, Tokyo, Japan, Nov. 1999.

Mat. Res. Soc. Symp. Vol. 613 © 2000 Materials Research Society

Planarization of Copper Damascene Interconnects by Spin-Etch Process: A Chemical Approach

Shyama P. Mukherjee, Joseph A. Levert
Honeywell Electronic Materials, 1349 Moffett Park Dr., Sunnyvale, CA, USA
Donald S. DeBear,
SEZ America Inc, 4824 South 40th St., Phoenix, AZ 85040, USA

ABSTRACT

The present work describes the process principles of "Spin-Etch Planarization" (SEP), an emerging method of planarization of dual damascene copper interconnects. The process involves a uniform removal of copper and the planarization of surface topography of copper interconnects by dispensing abrasive free etchants to a rotating wafer. The primary process parameters comprise of (a) Physics and chemistry of etchants, and (b) Nature of fluid flow on a spinning wafer. It is evident, that unlike conventional chemical-mechanical planarization, which has a large number of variables due to the presence of pads, normal load, and abrasives, SEP has a smaller number of process parameters and most of them are primary in nature. Based on our preliminary works, we have presented the basic technical parameters that contribute to the process and satisfy the basic requirements of planarization such as (a) Uniformity of removal (b) Removal rate (c) Degree of Planarization (d) Selectivity. The anticipated advantages and some inherent limitations are discussed in the context of process principles. We believe that when fully developed, SEP will be a simple, predictable and controllable process.

INTRODUCTION

Copper dual damascene architectures are increasingly being used for interconnects in integrated circuits. The copper is electrodeposited with processes that generate a nonplanar surface [1]. The current method for the planarization of copper dual damascene features and their underlying barrier metals is Chemical-Mechanical Planarization (CMP) process [2]. In CMP processes, the mechanical forces play a dominant role. The present work describes an emerging method of planarization called "Spin-Etch Planarization" (SEP). This is a chemical approach involving no mechanical force. The process involves the uniform removal as well as planarization of copper surface topography by dispensing abrasive free etchants onto a rotating wafer using a commercial spin-etch tool (SEZ 203) [3,4]. In this context, it is worth referring another chemical approach of planarization called Electrochemical Planarization based on the principle of anodic leveling or electropolishing [5]. This process also does not have applied mechanical forces. We believe that the advent of copper damascene interconnects and the polymeric and nanoporous low-k dielectric are the driving force for chemical approaches.

The removal process, of metal as used in SEP, can be described as "controlled" wet chemical dissolution of a metal surface in electrolytes on a rotating wafer. The term "etch " here should be expressed as a controlled wet "chemical polishing" rather than "etching" which does not fulfill the requirement of smooth surface finish. The wet chemical etching of metals in electrolytes is an electrochemical process. A large number of galvanic cells are created on the metal surfaces after immersion in electrolytes. Hence, the surface of a metal may be regarded as a complex multi-electrode system. The metal surface finish after wet chemical removal is controlled by the relative potential difference between the anodic and cathodic regions at a particular time, viscosity and local current density [6]. Another aspect of the chemical polishing is the formation of a viscous layer on the metal surface whose thickness in cavities is greater than on projections and edges and as a result, the dissolution rate of projections is higher and leads to leveling or planarization. The nature and the dimension of the viscous boundary layer developed during polishing play important roles in the SEP process.

OBJECTIVES

We have already reported some preliminary experimental results of SEP of copper interconnect [3,4]. The objectives of this paper are:
1) To identify the key basic technical principles and parameters that play a role in satisfying the primary requirements of planarization such as (a) the uniform removal of copper (b) the removal rate (c) the selectivity of copper removal (d) the degree of planarization, and
2) To evaluate qualitatively the processing advantages and limitations in the context of processing principles and characteristic features of the tool.

SPIN-ETCH PLANARIZATION TOOL

The tool (SEZ Spin Processor 203) is a commercially available single-wafer etch system. Spin etch tools are currently used for silicon wafer backside post-grind etch and for removing metal contamination from the wafers [7]. In this work we will emphasize the key elements of the tool that play important roles in the SEP process. The details have been described elsewhere [3, 4]. The key features of the tool providing simplicity and flexibility are as follows:
1) Capability of spinning 200 mm wafers on a contamination free nitrogen cushion at a wide range of rotation rates.
2) A programmable chemistry dispense arm that can deliver fluids at a selectable flow-rate and over any desired dispense profile sweeping radially across the wafer's surface, allowing for flexibility in controlling removal rate non-uniformity or for correcting for as-deposited copper thickness non-uniformities.
3) Three separate chemistry sources and dedicated process chambers for different etchants used during subsequent phases of the planarization process, and an additional chamber for cleaning/rinsing and drying the wafer.

An additional benefit of the SEP process and the Spin Processing system is its inherent low cost of ownership. The requirement of consumable materials is less; and because the SEZ 203 has integrated wafer cleaning and drying, no additional post-process cleaning equipment are required as part of the process flow. Fast etch rates and in-situ cleaning and drying will yield short process times, maintaining a high wafer throughput. Costs of ownership factors include waste disposal which is also lessened because all effluent is aqueous and contains no suspended particles. Relative ease of neutralization and the potential for recovery and re-use may further reduce costs.

SPIN ETCH PLANARIZATION PROCESS

A proposed flow of process steps for accomplishing copper dual damascene planarization is as follows:

Phase 1. A uniform removal and planarization of excess copper to the barrier layer interface with a single etchant that removes copper selectively without any dissolution of barrier layer such as Ta/TaN. Our present etchant system developed for copper planarization has no reactivity with either Ta or TaN.

Phase 2. The phase 2 involves a selective passivation of copper and a subsequent removal of the barrier layer, with a second step, by a second etchant that has a high selectivity to the barrier material.

EXPERIMENTAL RESULTS

The following typical experimental data is presented to help illustrate the basic principles as well as the connection between these principles and the controlling SEP process parameters. Pattern and blanket electroplated 200 mm copper wafers were used to obtain this data using Phase 1 SEP (described above).

The average blanket copper wafer removal rate was 14,000. Å/min with a 3σ non-uniformity of 9.17%. The copper removal rate was typically a function of the etchant chemistry and the spin speed while the uniformity was primarily a function of the etchant dispenses profile. The surface texture of plated copper before and after SEP was observed using a scanning electron microscopy (SEM). The SEM photomicrographs of copper surfaces before and after polishing are shown in Figure 1. The post-SEP surface was visually shiny and was smoother when measured by stylus profilometry (KLA-Tencor HRP-220). The post-SEP surface was 81. Å. (root mean square roughness) smoother than the as deposited copper. The post-SEP surface roughness was typically a function of the etchant chemistry as well as the spin speed.

We have evaluated the planarization performance by monitoring the change of feature recesses or protrusions near the end of Phase 1. Planarity is illustrated by the Degree of Planarization (DoP) by the following expression:

$$\text{DoP } (\%) = 100 \ (1 - R_f/R_i)$$

Where: R_i = initial recess before etching, (Å), and
R_f = final recess after SEP (Å).

Where 100% indicates no copper feature recess/dishing after SEP, while 0% indicates that all of

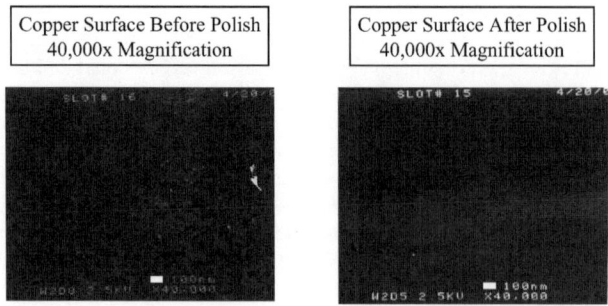

Figure 1. *SEM photomicrographs of Copper surfaces: (left) before Spin-Etch Planarization, (right) after Spin-Etch Planarization.*

the copper was removed from the bottom of the etched feature. The typical planarization for small features was a primarily a function of the etchant selection is given in the table below after 85% of the as deposited copper was removed.

Feature Width, μm	Initial Recess (Protrusion), Å	DoP, %
1.50	2,740.	81.
2.50	3,262.	78.

Erosion, of the dielectric spaces between dense small features, was evaluated using stylus profilometry (KLA-Tencor HRP-220) of a pattern electroplated copper wafer. The copper was removed from the field with a minor over-etch of the copper features (<500 Å recess/dishing) prior to profilometry. The resulting trace in Figure 2 shows no erosion. Erosion was only a function of the etchant chemistry.

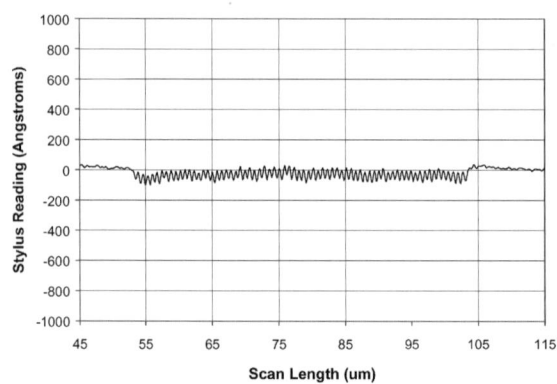

Figure 2. *Profilometry trace showing the absence of dielectric erosion after Spin Etch Planarization of dense featured patterned copper wafer surface.*

DISCUSSION

In the light of initial experimental results and from information from the existing literature, we anticipate that the following two key technical parameters play major roles in controlling SEP.

1) Physics and Chemistry of Etchants:

The composition and chemistry of the etchants determines the oxidizing power and dissolution behavior of the etchant and the selectivity of dissolution to different layers (such as copper layer, barrier layer and dielectric layer). The etchant used for this work was selective to copper only. The elimination of dielectric erosion is important for the selection of etchant for the barrier removal. The viscosity and the chemistry of the etchants controls the thickness and the nature of the diffusion boundary layer produced during the etching process as a function of spin speed.

2) Nature of fluid flow on a spinning wafer with non-planar topography:

This fluid flow phenomenon, with a particular etchant is controlled by the following key process parameters: (a) Spin Speed, (b) Flow Rate, and (c) Etchant Dispense Profile. The selected etchant contributes to the kinematic viscosity and chemical nature of interfacial layer developed during the etching process.

Based on the above information and from our experimental results it is evident that the chemistry and physics of the etchant was the dominant parameter for the copper removal rate, surface finish, and DoP. Spin speed was a secondary parameter, which affected the copper removal rate, surface finish, and DoP. Our experimental results on the surface roughness show that the surface roughness increases with spin speed. This is an indication that the spin speed might affect the viscous boundary layer, which consequently controls roughness and planarization. However, with a particular etchant ,the uniformity of removal is controlled primarily by the dispense profile which influences the distribution of the etchant across the wafer.

We anticipate that a copper diffusion boundary develops during the spin-etch process. The following relationship has been found to hold for the diffusion boundary layer representing mass transport on a rotating disk electrode process of metal removal:

$$\delta_d = D^{1/3} \upsilon^{1/6} \omega^{-1/2}$$

Where: δ_d = Diffusion boundary layer thickness,
D = Diffusion coefficient of Cu^{+2} ions,
υ = Kinematic viscosity,
ω = Rotation rate.

We believe that this relationship may be qualitatively applicable to the boundary layer thickness developed in SEP. The chemical nature and thickness of the diffusion boundary layer play key roles in controlling the SEP.

SEP performance was most affected by three process parameters: Etchant Chemistry, Spin speed, and Dispense profile. These three parameters are analogous to Slurry chemistry,

Platen speed, and Polishing head motion profiles of mechanically based polishing processes. However, mechanically based polishing is also affected by additional critical process parameters such as Normal load, and Pad material properties which consequently have complex secondary variables such as contact pressure, and pad conditioning. Although the SEP process has not yet been developed to the maturity of conventional planarization processes , SEP does have an advantage of a smaller process parameter set requiring control. The small SEP process parameter set is governed by two basic technical factors: the chemistry of etchants, and nature of fluid flow on a patterned wafer, which in turn are governed by well-known principles of chemistry and physics.

The absence of mechanical contact as well as the selectivity of abrasive free etchants allows SEP to eliminate the problems of dielectric erosion and surface scratching and abrasive particle contamination. The elimination erosion is an important requirement for the electrical performance of future high-density devices having polymeric or porous low-k dielectrics.

CONCLUSIONS

The SEP process has an intrinsic ability to impart high surface finish (Figure 1) and planarize copper surfaces. SEP is a controlled wet chemical removal process of copper involving an electrochemical dissolution process. Dissolution reactions are modified by the diffusion of chemically reactive species within a diffusion boundary layer modified by etchant flow on the rotating wafer. The relatively small number of SEP process parameters has the prospect of yielding a simpler process based on previously codified principles of chemistry and physics.

ACKNOWLEDGMENTS

The authors want to thank Lynn Forester and Michael Thomas of Honeywell Electronic Materials STAR Center and Michael West of SEZ for their support.

REFERENCES

1. P. C. Andricacos, C. Uzoh, J. O. Dukovic, J. Horkins, and H. Deligianni, IBM J. Res. Development **42** (5) 567-574 (1998).
2. P. Wrschka, J. Hernandez, G. S. Oehrlein and J. King, J. Electrochemical Soc. **147** (2) 206-712 (2000).
3. J. A. Levert, S. P. Mukherjee, D. S. DeBear, Semicon Japan 99: SEMI Technology Symposium, **9** 4-73 to 4-81 (1999).
4. D. S. DeBear, J. A. Levert, S. P. Mukherjee, Solid State Technology, March, 53-58.(2000).
5. R. J. Contolini, S. T. Mayer, R. T. Graf, L. Tarte, A. F. Bernhardt, Solid State Technology, June 155-158 (1997).
6. J. W. Bloor, J. Metals of Australia **4**, 276-282 (1972).
7. P. S. Lysaght, and M. West, Solid State Technology, November 63-70 (1999).

AUTHOR INDEX

SUBJECT INDEX